Springer Series on Touch and Haptic Systems

The Springer Series on Touch and Haptic Systems is published in collaboration with the EuroHaptics Society. It is focused on publishing new advances and developments in all aspects of haptics. Haptics is a multi-disciplinary field with researchers from Psychology, Physiology, Neurology, Engineering, and Computer Science (amongst others) contributing to a better understanding of the sense of touch, and researching into how to improve and reproduce Haptic interaction artificially in order to simulate real scenarios. The series includes monographs focused on specific topics, edited volumes covering general topics from different perspectives, and selected PhD theses. Books in this series focus on Haptics or Haptic interfaces including:

- Neuroscience
- Perception & Psychophysics
- Haptic Rendering
- Devices and Technology
- Medical & Rehabilitation Applications
- Art & Design Applications
- Collaborative Haptics
- Mutimodal Interaction
- Tactile Display & Tactiles
- Sensing
- Haptic HCI (Interaction, Visualization)

Bingxu Li • Tyler Cody

Design Principles for Integrated Haptic Systems

Springer

Bingxu Li
Smarthap LLC
Charlottesville, VA, USA

Tyler Cody
Smarthap LLC
Charlottesville, VA, USA

ISSN 2192-2977 ISSN 2192-2985 (electronic)
Springer Series on Touch and Haptic Systems
ISBN 978-3-032-16985-3 ISBN 978-3-032-16986-0 (eBook)
https://doi.org/10.1007/978-3-032-16986-0

This Springer imprint is published by the registered company Springer Nature Switzerland AG
The registered company address is: Gewerbestrasse 11, 6330 Cham, Switzerland

Preface

Touch is the first sense through which we learn the world—and the last frontier of artificial perception, precisely because it is inseparable from the body itself. Unlike audition, which is largely passive, or vision, which—though active—is spatially centralized and cortically filtered, haptics is distributed, reciprocal, and grounded in mechanics. The human haptic system does not merely sense; it filters, re-weights, and infers meaning through the physics of skin and motion. This bio-inspired perspective forms the central thesis of this book: any artificial system aspiring to perceive, interpret, or communicate through touch must be built upon the same integrated logic that governs human haptics—a system in which sensing, inference, and control are inseparable.

Drawing from over four decades of haptics research and more than 300 works across the scientific and engineering literature, this book proposes a new worldview—integrative haptics—as both a design philosophy and a disciplinary framework. Rooted in systems engineering, it treats haptics not as a collection of sensors or devices but as a closed-loop system that connects biomechanics, computation, and application into coherent architectures. This systems-level synthesis establishes a foundation for future generations of haptic technologies that are informed by, but not limited to, biological precedent—technologies that evolve from imitating human touch to achieving tactile perceptual intelligence of their own.

Readers will find here both a vision and a practical framework: a vision that positions touch as the next frontier of embodied intelligence, guided by systems integration, and a framework that translates the science of haptics into actionable principles for design. For the scientist, the book connects perception and mechanics through first principles; for the designer, it provides a new lens for shaping systems that can both feel and understand. Taken together, these perspectives define the aim of this work—to advance touch from sensation to system, and from system to understanding.

Charlottesville, VA, USA
October 2025

Bingxu Li
Tyler Cody

Declarations

Competing Interests The authors have no competing interests to declare that are relevant to the content of this manuscript.

Contents

About the Authors

Dr. Bingxu Li is Chief Technology Officer at SmartHap, a haptic technology company based in Virginia, USA. She received her Ph.D. in Systems Engineering in 2023 from the University of Virginia, where she researched haptic science, focusing on the relationship between skin mechanics and psychophysics as well as how that relationship varies across individuals. Her current research focuses on artificial skin, including sensing and actuation mechanisms for recreating the human sense of touch. She has received significant recognition for her first-authored work, including the IEEE Transactions on Haptics 2024 Best Paper Award, a cover mention in the *Journal of Physiology*, and expositions on her research in University of Virginia's UVA Daily News and the global publisher ATLAS News, among other best paper awards.

Dr. Tyler Cody is Chief Executive Officer at SmartHap, as well as Associate Professor of Data Science at the University of Virginia in Charlottesville, Virginia, USA, where he serves as Director of the Artificial General Intelligence Lab. He received his Ph.D. in Systems Engineering in 2021 from the University of Virginia. His research interests include advancing the systems engineering discipline with formalism and first principles that address the novel systems challenges introduced by artificial intelligence, a research area where haptic technology provides a canonical domain. His research has received several best paper awards, including a career publication award from the International Test and Evaluation Association (ITEA) in 2024 for his work on best practices for testing machine learning.

Chapter 1
System Design of Integrated Haptic Systems

Abstract Imagine navigating the world without the ability to feel; a world where the warmth of a loved one's hand, the softness of a pillow, or the firmness of a tool in your grip is lost to you. Our sense of touch is more than just a background sensation; it is an intimate, dynamic dialogue between the body and the environment, shaping how we perceive and interact with everything around us. As essential as vision and hearing, touch allows us to experience the world in high definition—offering immediate, tangible feedback that guides our movements, protects us from harm, and connects us to others in deeply meaningful ways.

1.1 Haptic Technology Design Needs a Systems Mindset

Touch is not passive. It is a richly textured, real-time interaction between body and world. In building devices and machines that can "feel," haptic system designers are not merely engineering sensors and actuators; they are reconstructing an intricate dialogue between mechanics, perception, and control. Future haptic technology capable of recreating touch with lifelike realism demands this very intricacy. Today's practitioners, meanwhile, must translate haptic science into technologies that meet application needs and create distinctive value.

Systems engineering offers a framework to systematically approach this dialogue from first principles. Haptics provides grounding in the physics of contact and the neurophysiology of perception, while systems engineering situates these dynamics within interdependent components and a process aimed at achieving design goals. In this book, we propose a haptic system concept and associated design process guided by first principles that model haptic technology as an integrated system. Our contribution is a step toward a design philosophy for *integrated haptic systems*—one that treats them as closed, integrated systems in which software, hardware, and end-use application shape and constrain one another. Our adaptable concepts provide a framing for this book's broader discourse on principles for haptic design.

B. Li, T. Cody, *Design Principles for Integrated Haptic Systems*, Springer Series on Touch and Haptic Systems, https://doi.org/10.1007/978-3-032-16986-0_1

1.1.1 What Is a System

In systems theory and science, a system is broadly defined as a set of interrelated components that interact to fulfill a common purpose [4]. Building on this perspective, systems engineering and design emphasize the formulation of goals, their decomposition into subsystems and components, and the integration of these components into systems that realize those goals [22]. System design goals are the desired objectives or qualities that a system must achieve within its context. Such goals may address both the characteristics of the system itself and the outcomes of its operation. Grounded in stakeholder needs and contextual factors, system design goals are refined into subsystem objectives that guide the design of individual components. An effective system design aligns component functions with system-level objectives, ensuring that the whole meets its intended purpose and performance requirements.

Definition 1.1 A **system** is an organized set of interrelated components that interact within defined boundaries to achieve a specific purpose or function.

Definition 1.2 A **system design goal** is an intended objective or desired quality that guides the development of a system to ensure it meets functional requirements and performs effectively within its context.

1.1.2 What Are Haptic Systems

To take a systems approach to haptic design, we need to develop a haptic system concept, and for that, a general concept of haptics. Haptics concerns the science and engineering of touch. It includes both the study of how humans perceive the world through physical contact—by way of the skin, muscles, joints and tendons—and the development of technologies that can recreate or respond to touch—through forces, vibrations, and motions. Across disciplines, the term haptics has been defined in ways that reflect the priorities and contexts of each field [6].

In psychology and neuroscience, haptics is primarily concerned with the mechanisms underlying tactile perception, which arise from the integration of cutaneous (arising from skin deformation) and kinesthetic (originating in muscles, tendons, and joints) feedback. Together, these sensory pathways provide the basis for perceiving object properties such as shape, compliance, and texture [15, 26]. In human-computer interaction (HCI) and virtual reality, haptics is typically defined in terms of interface methods that apply mechanical stimuli—such as forces, vibrations, and motions—to enable users to interact with virtual objects or remotely controlled machines [23]. In robotics, haptics is commonly understood as the measurement and control of physical interactions between robots and their environment, where touch provides critical information for manipulation and safety [18, 20]. Even in philosophy and cognitive science, haptics has been discussed as a key modality of embodied cognition, underscoring its unique role in grounding perception in physical interaction [19].

Haptics focuses on two areas: how touch enables perception in humans, and how engineered systems can replicate or extend this capability. Drawing from these different fields, we can treat haptics as the science and engineering of the sense of touch—both natural and mediated—encompassing real or simulated interactions between humans (or robots) and environments (physical or virtual). It involves sensory pathways (e.g., cutaneous, kinesthetic, and proprioceptive) and action (e.g., exploration, manipulation) to perceive, interpret, and respond to object and spatial properties. This concept of haptics emphasizes its structural, behavioral, and functional nature—those systems aspects we can use to frame design.

Definition 1.3 **Haptics** is the science and engineering of the sense of touch involving sensing and action to perceive, interpret, and respond to object and spatial properties.

While haptics is defined across domains, the concept of haptic system has not been standardized in the literature. In current practice, haptics is often approached from a component-level perspective—with design scope defined in terms of specific technologies such as tactile sensors, vibrotactile actuators, or force-feedback devices, and with the goal of achieving narrow functionality such as pressure sensing, vibration rendering, or kinesthetic feedback. As a result, haptics is treated as a set of discrete capabilities rather than as an integrated modality requiring a cohesive systems design framework.

To address this, we adopt a systems engineering perspective [4, 22]. A haptic system is a system that has the functional role of delivering haptic capability within a larger architecture. Under Definition 1.3, that capability may include sensing, simulating, or reproducing touch-related cues.

Definition 1.4 A **haptic system** is a functional system composed of interacting components that work together to deliver haptic capability, to support perception and interaction within a larger physical or digital architecture.

1.2 Integrated Haptics and Closed-Loop Applications

A systems mindset in haptic design naturally leads to the recognition that touch itself is a closed-loop phenomenon. Touch involves reciprocal sensing and action—each informing and constraining the other. Recognizing the fundamentally closed-loop nature of haptic phenomena motivates a structured, integrated approach to haptic system design. Because touch arises from the continuous coupling of sensing and action, any engineered system that seeks to deliver haptic capability must coordinate across the physical (hardware), computational (software), and behavioral (application) domains. Section 1.2 develops this integrated systems perspective, beginning in Sect. 1.2.1 with the argument that closure is not optional but intrinsic to haptics itself. The sections that follow build on this foundation to propose the concept of an integrated haptic system and its corresponding integrated design process, culminating in the special case of inference-centric haptic systems.

1.2.1 Integration and Closure as Fundamental to Haptics

In most areas of engineering, closed-loop control is the cornerstone of robust system performance. Sensors provide feedback, actuators adjust their outputs, and the system continually corrects itself to meet its target state. Yet, closed-loop coordination remains a major limitation in haptic systems. Most current devices operate with open-loop strategies: they maximize component-level performance—such as increasing sensor resolution, actuator bandwidth, or rendering fidelity—while neglecting how the system as a whole should adapt to the user and the environment to achieve the objective goal, see Fig. 1.1.

This lack of closed-loop integration poses a major information bottleneck—especially as we seek to embed haptic systems within larger, more complex systems, such as humanoid robots operating in dynamic, real-world environments. In such scenarios, touch is not a standalone modality. It becomes one of many sensory inputs that must be interpreted and acted upon in coordination with other modalities such as vision, speech, and environmental context. Such robots must function in varied and unpredictable settings—grasping delicate objects, responding to human touch, or navigating crowded spaces. A robot assisting an elderly user in a home, or interacting with a child in a classroom, must be able to modulate its tactile behavior—adjusting how much, when to use, and what kind of touch—depending on the situation.

In this context, more data or higher fidelity at the component-level does not guarantee better system performance. A haptic system must not only sense, but also perceive what information matters, when it matters, and why. Excessive or poorly prioritized tactile data can overwhelm interaction pipelines, delay responses, and destabilize behavior—particularly in time-critical or safety-sensitive settings.

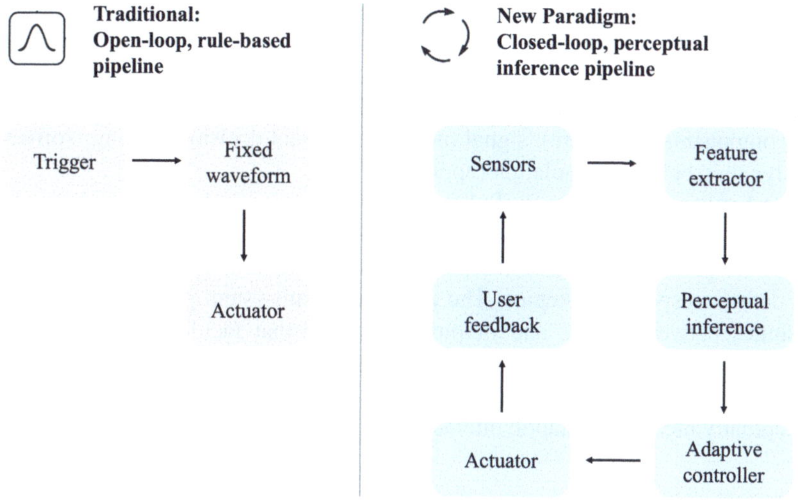

Fig. 1.1 Comparison between an open-loop haptic system (left) and an inference-centric closed-loop system (right). The closed-loop design integrates sensing, perceptual inference, and adaptive control to continuously adjust actuation based on feedback, enabling perception-aligned and context-adaptive behavior

Just as humans filter sensory input for relevance, intelligent systems must regulate information flow to maintain responsiveness and stability.

Haptic capability, therefore, emerges from closed-loop integration rather than isolated optimization. Sensors, actuators, and software algorithms must operate as parts of a unified architecture that prioritizes data according to context and higher-level goals. Sensors should capture perceptually and operationally meaningful features; actuators should emphasize timing, precision, and interpretability; and software algorithms should infer, adapt, and anticipate to align system behavior with user intent and task outcomes. Such integration transforms haptic devices from high-fidelity instruments into intelligent systems capable of modulating their tactile behavior dynamically, allocating computational and sensory resources where they are most effective.

1.2.2 Integrated Haptic Systems

To advance haptic design philosophy toward this closed systems paradigm, we propose *integrated haptic systems*, a form of haptic system in which the delivery of haptic capability is achieved through the structured coordination of three subsystems: hardware, software, and application. These subsystems form closed-loop interaction between the haptic system and its functional use in applications, such as in control or decision-making.

Integrated haptic systems are designed from the top down to satisfy system-level design goals. Delivering haptic capabilities at the systems level requires that high-level objectives (e.g., initiate a firm, trustworthy handshake) be translated into coherent subsystem-level design goals. These goals are then allocated across physical components (hardware), signal interpretation and decision-making (software), and dynamic behavior regulation (application).

The hardware subsystem includes sensors, actuators, and mechanical structures that enable physical interaction with the environment or the user. The software subsystem processes signals, interprets sensory data, encodes perceptual models, and defines response strategies. The application subsystem governs the feedback dynamics between inputs and outputs—ensuring that tactile cues are delivered accurately and effectively in real-time. Together, these three subsystems form the foundation of a closed-loop architecture, which is essential for adaptive and perceptually meaningful haptic interaction.

Definition 1.5 An **integrated haptic system** is a haptic system composed of three coordinated subsystems—hardware, software, and application—engineered to deliver haptic capabilities through closed-loop interaction.

1.2.3 Integrated Haptic System Design Process

If an integrated haptic system is defined by the coordination of hardware, software, and application subsystems to deliver closed-loop haptic capability, then its design must likewise be an integrated process, see Fig. 1.2. The integrated haptic system design process provides a simple, systematic means to translate system-level intent into coherent subsystem designs grounded in first principles. This process connects the "why" of design, the system's purpose and context, to the "how" of realizing that purpose through interaction among subsystems.

The process proceeds through four sequential steps.

- First, given overall system design goals, the designer identifies the haptic system design goals—the specific capabilities or experiences that the haptic system must enable within the larger system. These goals define the contribution of haptics to system-level performance.
- Second, the designer identifies the first principles relevant to these goals. These could include physical laws, perceptual mechanisms, or psychophysical relationships that determine how haptic phenomena can be achieved or constrained.
- Third, the haptic system design goals are decomposed across the three subsystems of the integrated haptic system—hardware, software, and application—guided by those first principles. This decomposition ensures that subsystem functions collectively realize the intended closed-loop behavior.

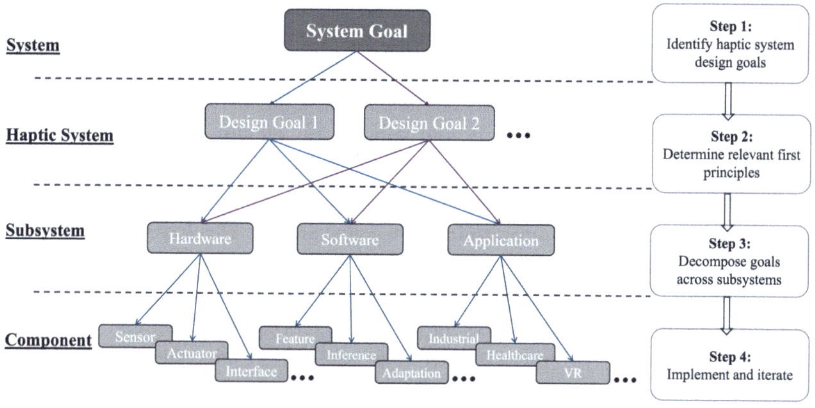

Fig. 1.2 System structure of the integrated haptic system and its four-step design process. The diagram links system-level goals to hardware, software, and application subsystems through goal decomposition and iterative refinement, illustrating how the integrated haptic system design process aligns perception, function, and control

- Fourth, the design is implemented and iterated. Development proceeds in cycles of modeling, prototyping, and evaluation, refining subsystem interactions until system-level objectives are satisfied.

This stepwise process unites systems engineering discipline with haptic science in the practice of design. It provides a traceable path from goals to principles to architecture to realization, ensuring that integrated haptic systems are not merely assembled from components but engineered as coherent wholes.

Definition 1.6 The **integrated haptic system design process** is a four-step method for designing integrated haptic systems: (1) identify haptic system design goals from system design goals; (2) identify relevant first principles; (3) decompose goals across hardware, software, and application subsystems guided by those principles; and (4) design, develop, and iterate to achieve system-level goals.

1.2.4 Inference-Centric Haptic Systems

In this book, we follow the integrated haptic system design process to explore and demonstrate the use of haptic science in design. Our subject of design is a new class of integrated haptic systems that is both bio-inspired as well as representative of future haptic technologies—inference-centric haptic systems. Future haptic systems will not be defined solely by their ability to sense, transmit, or recreate mechanical interaction, but by their capacity to support inference—the transformation of raw tactile signals into meaningful perceptual information that guides action and decision-making within their larger systems. This evolution reframes haptic systems from feedback instruments to intelligent subsystems capable of adapting behavior dynamically to context, user intent, and environmental conditions.

In the integrated haptic systems framework, inference emerges from coordination across hardware, software, and application subsystems. The hardware subsystem provides structured, information-rich signals reflecting contact dynamics and material interaction. The software subsystem encodes perceptual and physical models that transform these signals into higher-order representations, e.g., estimating properties such as force, intent, or texture. The application subsystem then uses these inferred states to modulate actuation, allocate computational resources, and shape interactive behavior. Designing for inference focuses the integrated haptic system design process on first principles that describe not only physical relationships but also the information transformations that support perceptual inference.

Definition 1.7 An **inference-centric haptic system** is an integrated haptic system in which coordination across hardware, software, and application subsystems enables the transformation of tactile signals into perceptual inferences that guide action and decision-making.

1.3 Integration and Inference in Biological Haptic Systems

Adopting an integrated haptic system framework has a direct analogy to the human body, which exemplifies a fully integrated, closed-loop sensory-control architecture, see Fig. 1.3. The human sense of touch is more than a collection of raw sensory data; it is a coordinated system where sensing, inferencing, and action are intricately coupled to work together toward adaptive, context-sensitive goals.

1.3.1 Integration and Control in Biological Haptics

When the finger contacts an object, tactile signals are generated at the skin through deformation patterns. Mechanoreceptors transduce these patterns into electrical impulses that travel along afferent fibers to the nervous system, encoding spatial, temporal, and dynamic information about physical contact, see Fig. 1.4. What makes the somatosensory system remarkable, however, is not just the richness of these

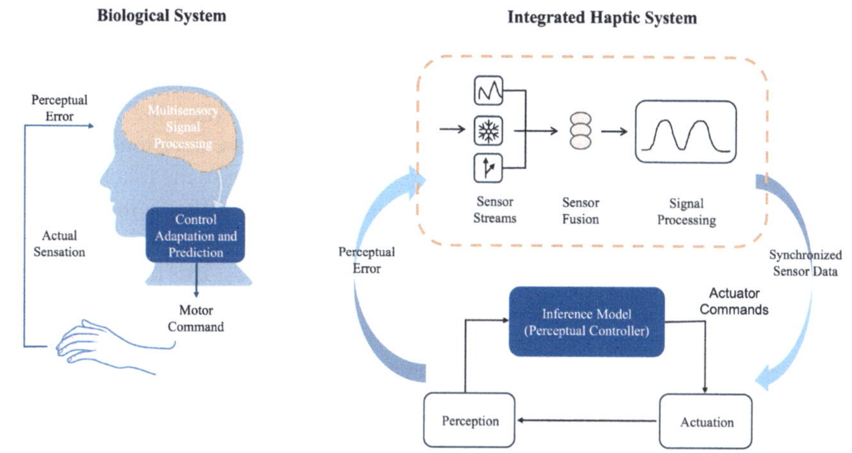

Fig. 1.3 Analogy between the human sensorimotor loop and an inference-centric haptic system. Both use closed-loop inference—linking sensing, perception, and action—to achieve adaptive, context-aware control

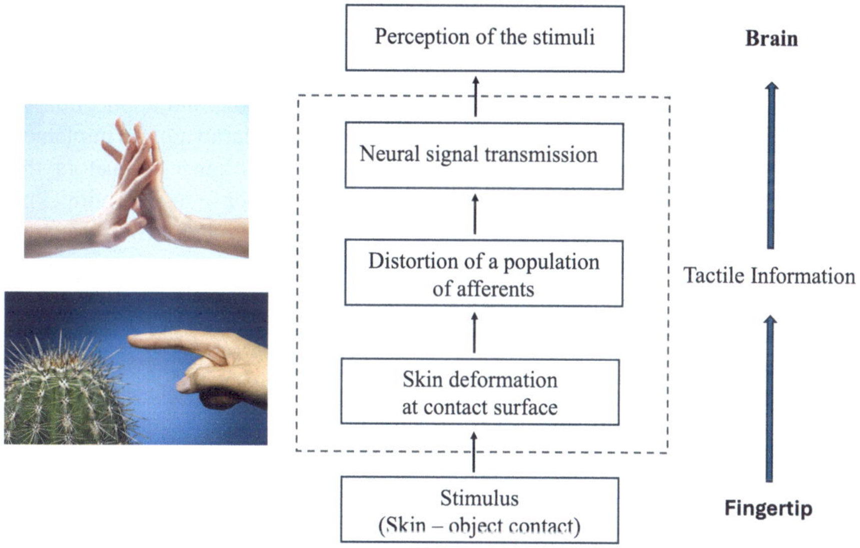

Fig. 1.4 Tactile information transmission—from skin to somatosensory area of brain

inputs, it is how they are processed, interpreted, and acted upon in layered feedback loops, each tuned to different timescales and functions [24, 25].

In biological systems, tactile signals from mechanoreceptors are continuously transmitted as high-resolution afferent input to multiple neural centers, such as the spinal cord, cerebellum, and cortex. The spinal cord governs low-latency reflex arcs that enable rapid, local responses to sudden events, such as adjusting grip force in response to slippage, operating below conscious awareness to ensure immediate safety and stability [2, 11, 13]. The cerebellum serves as a predictive and corrective layer, comparing incoming tactile information with internal model to anticipate and refine motor outputs [7, 28]. Higher-level processing units (i.e., the cortex), by contrast, are responsible for making deliberate, context-aware decisions—judging whether an object is sharp, identifying textures, or interpreting social touch [14, 16].

Importantly, these mechanisms do not operate independently or in sequence. Instead, they draw from shared sensory input by selectively processing, filtering, and prioritizing information depending on the task and goal [3, 12]. This distributed processing prevents overload and supports adaptive information routing—the nervous system does not always rely on the highest resolution of sensory input, but instead emphasizes relevant cues (e.g., vibration when exploring texture) while suppressing irrelevant ones.

This biological organization offers a direct analogy for engineering integrated haptic systems that can serve higher-level system goals. Just as the body coordinates between peripheral encoding (mechanoreceptors transduce physical contact into neural signals), neural processing (where intermediate centers such as the spinal cord and cerebellum filter, predict, and refine those signals), and command-action control (where higher cortical centers transform sensory information into task-directed motor decisions). By analogy, given a system-level goal, an integrated haptic system distributes these roles across its three subsystems: hardware for sensing and actuation, software for modeling and prediction, and application for decision, coordination, and end-use in the broader system. Hardware can implement fast, local controllers that act on salient signals—for instance, actuators that instantly adjust grip pressure or vibration amplitude in response to slip, mimicking spinal reflexes. Software modules can implement prediction and error correction, aligning perceptual states with task-defined targets to adjust actuation strategies in a closed loop to minimize perceptual error—an analogue to the cerebellum's role in error correction and anticipatory control. Higher-level controllers in the application can interpret perceptual states in the context of broader system goals, supporting more abstract decision-making such as object categorization or safe navigation in unstructured environments.

Notably, control in this framework is not confined to the internal loop of the integrated haptic system itself. While the hardware, software, and application subsystems regulate local feedback and interaction, the application subsystem can also operate at the level of the larger system architecture, aligning haptic outputs with higher-level goals. In this way, the integrated haptic system framework supports both subsystem-level regulation and whole-system coordination, mirroring the layered organization of the human somatosensory system.

Crucially, an integrated haptic system should not overwhelm the system with all available data. Like the human nervous system, it should intelligently prioritize information based on current tasks and system-level goals. For example, a robot executing a low-risk motion may only require kinesthetic feedback, whereas manipulating fragile or irregular objects demands richer cutaneous cues. In high-stress or time-critical contexts, excessive sensory data can act as noise. The brain manages this by suppressing irrelevant inputs and amplifying meaningful ones [5, 10, 17]; engineered haptic systems must do the same.

This biological analogy also reinforces the principle of integration and closed-loop control. Our skin are not just sensors, our muscles are not just actuators, and our brain is not just a controller. They are interdependent and co-evolved components of a system that develops, operates, and adapts together to achieve shared goals. Likewise, the performance of a haptic system does not emerge from optimizing any single subsystem in isolation, but from the coordination of all subsystems in service of system-level objectives.

1.3.2 Inference as an Emerging Lens in Biological Haptics

The inference-centric haptic system concept mirrors an inference-centric framing of the human sensorimotor system, which is increasingly understood not as a command-driven controller but as an inference system. Under such a framework, the nervous system does not simply react to inputs but continuously generates predictions about sensory outcomes, compares them to incoming signals, and minimizes the discrepancy between expected and actual states [1, 9, 21]. Sensory signals are not treated as isolated facts, but as evidence for or against the brain's current hypotheses about the state of the world. Movement itself—grasping, pressing, adjusting—is generated not as a feed-forward command, but as an action that fulfills sensory predictions. In this view, perception and action both serve to resolve uncertainty by reducing prediction error.

In the context of touch, this means that the sensorimotor system is constantly inferring the causes of skin deformation—predicting object compliance, friction, and shape—and adjusting exploratory movements to minimize discrepancies between expected and sensed cues. At the periphery, mechanoreceptors in the skin transduce deformation into neural signals that carry information about contact dynamics such as force, contact area, and vibration. These afferents provide the evidence against which predictions are tested [12]. At intermediate levels, such as the cerebellum, internal models are thought to generate forward predictions of sensory consequences of movement [8, 27], which can be updated when mismatches arise. At the highest levels, the cortex integrates tactile input with contextual and cognitive factors, allowing the system to infer not only physical properties (e.g., compliance, roughness) but also social meaning [14]. Across all of these stages, the brain's role is less to issue commands and more to infer the causes of sensory input and act in ways that reduce uncertainty [1, 9].

Similar principles can be applied to the design of inference-centric haptic systems. First, the hardware analogues (skin, mechanoreceptors, musculoskeletal tissues) provide rich, noisy sensory data that are most relevant for perception and need to be encoded with fidelity. Second, software analogues (neural integration, predictive coding, generative modeling) transform these raw data into perceptual states and expectations. Third, application analogues (motor commands and feedback loops) regulate behavior to minimize discrepancies between predicted and observed outcomes. In this sense, the nervous system embodies an inference-centric architecture: a closed-loop where sensing, prediction, and action are co-optimized to support adaptive, context-sensitive system-level performance.

Likewise, an inference-centric haptic system should be conceived not as a device that simply measures forces or delivers fixed feedback patterns, but as an architecture in which hardware, software, and application subsystems collectively contribute to inference. By embedding predictive models and closed-loop control to the system, raw sensor data can be transformed into perceptually meaningful states, and actuation can be dynamically tuned to minimize perceptual error. This mirrors the biological counterpart that the nervous system's ultimate goal is not to reproduce mechanical signals per se, but to support reliable, contextually appropriate inferences about the environment.

1.4 Summary and Roadmap

This chapter has established the conceptual foundation of the book: a design philosophy that unites the physics and perception of touch with the systems engineering principles needed to reproduce them. We began by recognizing that haptics is inherently closed-loop—an interaction in which sensing and action continuously inform one another. Building on this premise, we defined the integrated haptic system and introduced its corresponding integrated design process, a structured method for translating system goals into coordinated subsystem architectures. Extending this framework, we proposed the concept of the inference-centric haptic system, which emphasizes that future haptic technologies must not only sense and act but also infer, transforming tactile data into perceptual intelligence that guides adaptive behavior.

The human body provides both validation and inspiration for this framework. Biological haptics demonstrates how sensing, inference, and control co-evolve across multiple layers of feedback, yielding a resilient and adaptive system. This analogy reinforces the idea that effective haptic design depends on integration across domains and coordination across scales, rather than optimization of isolated components.

The organization of this book mirrors the integrated design process itself. Chapters 2 and 3 examine the first principles of haptic science: the physical, physiological, and psychophysical laws that govern perception and interaction. These principles form the analytical basis for decomposing system-level goals. Chapters 4, 5, and 6 address the three subsystems of the integrated haptic system—hardware, software, and application—each explored in depth building to a case study: the design of an artificial skin, an inference-centric haptic system that models compliance and accounts for individual differences in psychophysical perception. Chapter 7 concludes with a synthesis of these ideas and a forward-looking discussion on the future of haptics and the next generations of tactile technologies.

The structure of the book therefore embodies the very philosophy it presents: beginning with goals and principles, decomposing them into interacting subsystems, and re-integrating them into a coherent whole. We now turn to the first principles that underlie all haptic phenomena and the foundation upon which every integrated haptic system must be built.

References

1. Adams, R.A., Shipp, S., Friston, K.J.: Predictions not commands: active inference in the motor system. Brain Struct. Funct. **218**(3), 611–643 (2013)
2. Cole, K.J., Abbs, J.H.: Grip force adjustments evoked by load force perturbations of a grasped object. J. Neurophysiol. **60**(4), 1513–1522 (1988)
3. Desmurget, M., Grafton, S.: Forward modeling allows feedback control for fast reaching movements. Trends Cogn. Sci. **4**(11), 423–431 (2000)
4. Dori, D., Sillitto, H., Griego, R.M., McKinney, D., Arnold, E.P., Godfrey, P., Martin, J., Jackson, S., Krob, D.: System definition, system worldviews, and systemness characteristics. IEEE Syst. J. **14**(2), 1538–1548 (2019)
5. Egner, T., Hirsch, J.: Cognitive control mechanisms resolve conflict through cortical amplification of task-relevant information. Nature Neurosci. **8**(12), 1784–1790 (2005)
6. El Saddik, A.: The potential of haptics technologies. IEEE Instr. Measur. Mag. **10**(1), 10–17 (2007)
7. Flanagan, J.R., Wing, A.M.: The role of internal models in motion planning and control: evidence from grip force adjustments during movements of hand-held loads. J. Neurosci. **17**(4), 1519–1528 (1997)
8. Flanagan, J.R., Vetter, P., Johansson, R.S., Wolpert, D.M.: Prediction precedes control in motor learning. Current Biol. **13**(2), 146–150 (2003)
9. Friston, K.: Learning and inference in the brain. Neural Netw. **16**(9), 1325–1352 (2003)
10. Gomez-Ramirez, M., Hysaj, K., Niebur, E.: Neural mechanisms of selective attention in the somatosensory system. J. Neurophysiol. **116**(3), 1218–1231 (2016)
11. Häger-Ross, C., Johansson, R.S.: Nondigital afferent input in reactive control of fingertip forces during precision grip. Exp. Brain Res. **110**(1), 131–141 (1996)
12. Johansson, R.S., Flanagan, J.R.: Coding and use of tactile signals from the fingertips in object manipulation tasks. Nature Rev. Neurosci. **10**(5), 345–359 (2009)
13. Johansson, R.S., Westling, G.: Roles of glabrous skin receptors and sensorimotor memory in automatic control of precision grip when lifting rougher or more slippery objects. Exp. Brain Res. **56**, 550–564 (1984)
14. Keysers, C., Kaas, J.H., Gazzola, V.: Somatosensation in social perception. Nature Rev. Neurosci. **11**(6), 417–428 (2010)
15. Lederman, S.J., Klatzky, R.L.: Haptic perception: a tutorial. Atten. Percept. Psychophys. **71**(7), 1439–1459 (2009)
16. Lieber, J.D., Bensmaia, S.J.: High-dimensional representation of texture in somatosensory cortex of primates. Proc. Natl. Acad. Sci. **116**(8), 3268–3277 (2019)
17. Macaluso, E., Driver, J.: Multisensory spatial interactions: a window onto functional integration in the human brain. Trends Neurosci. **28**(5), 264–271 (2005)
18. Van der Meijden, O.A., Schijven, M.P.: The value of haptic feedback in conventional and robot-assisted minimal invasive surgery and virtual reality training: a current review. Surgical Endoscopy **23**(6), 1180–1190 (2009)

19. Noë, A.: Action in Perception. MIT Press, Cambridge (2004)
20. Okamura, A.M.: Haptic feedback in robot-assisted minimally invasive surgery. Curr. Opin. Urol. **19**(1), 102–107 (2009)
21. Parr, T., Limanowski, J., Rawji, V., Friston, K.: The computational neurology of movement under active inference. Brain **144**(6), 1799–1818 (2021)
22. Salado, A.: A systems-theoretic articulation of stakeholder needs and system requirements. Syst. Eng. **24**(2), 83–99 (2021)
23. Salisbury, J.K., Srinivasan, M.A.: Phantom-based haptic interaction with virtual objects. IEEE Comput. Graph. Appl. **17**(5), 6–10 (2006)
24. Scott, S.H.: Optimal feedback control and the neural basis of volitional motor control. Nature Rev. Neurosci. **5**(7), 532–545 (2004)
25. Shadmehr, R., Krakauer, J.W.: A computational neuroanatomy for motor control. Exp. Brain Res. **185**, 359–381 (2008)
26. Srinivasan, M.A., LaMotte, R.H.: Tactual discrimination of softness. J. Neurophysiol. **73**(1), 88–101 (1995)
27. Wolpert, D.M., Ghahramani, Z., Jordan, M.I.: An internal model for sensorimotor integration. Science **269**(5232), 1880–1882 (1995)
28. Wolpert, D.M., Miall, R.C., Kawato, M.: Internal models in the cerebellum. Trends Cogn. Sci. **2**(9), 338–347 (1998)

Chapter 2
First Foundational Principle: Touch from Sensing to Perception

Abstract This chapter establishes the first foundational principle in the design of integrated haptic systems: the link between the mechanical behavior of human skin and the perceptual experiences it gives rise to. Human touch is fundamentally a process of mechanical stimulation interpreted by the nervous system. Every tactile experience begins with the deformation of the skin under contact. Understanding how these mechanical events are encoded and transformed into perceptual responses provides the scientific grounding for designing integrated haptic systems that can replicate or interface with the natural human sense of touch.

2.1 Skin Mechanics as the Gateway to Perception

Skin mechanics are the gateway to tactile perception. Mechanical inputs, such as indentation, stretch, or vibration, do not directly reach the brain; instead, they must first deform the skin, activate specific mechanoreceptors, and trigger neural coding processes that culminate in perceptual outcomes. This chain of events transforms a physical interaction into a felt experience.

Skin mechanics describe how the skin behaves as a physical material when subjected to external forces. The skin is not a uniform surface but a multilayered, viscoelastic, and anisotropic structure that deforms through stretching, compression, and shear in response to contact. Skin biomechanics extends this understanding by relating these mechanical responses to biological function, explaining how deformation patterns activate different classes of mechanoreceptors and initiate neural signaling pathways. Skin deformation, in turn, refers to the measurable, localized changes in the skin's geometry—its displacement, strain, or curvature—that occur during physical interaction and serve as the mechanical input to the sensory system.

By examining how biomechanical stimulation drives perceptual meaning, this chapter establishes the first scientific foundation for haptic system design. It explores how specific patterns of skin deformation lead to distinct psychophysical outcomes, providing a bridge between physical input and perceptual output. In doing so, it identifies the fundamental cues that underlie key tactile dimensions—

B. Li, T. Cody, *Design Principles for Integrated Haptic Systems*, Springer Series on Touch and Haptic Systems, https://doi.org/10.1007/978-3-032-16986-0_2

compliance, roughness, friction, and shape. By tracing the flow of tactile information from biomechanical input to neural encoding, we reveal how the physics of skin deformation underlies psychophysical performance and why this knowledge is essential for defining subsystem goals in integrated haptic system design.

Foundational Principle 1—Biomechanical–Informational Coupling
Tactile perception is both physical and informational. The physical arises from the biomechanical response of the skin to external forces; the informational arises from neural encoding and the pathways that generate psychophysical response. The relationship between biomechanics and psychophysics can be characterized through measurable mechanical cues that link physical deformation to perceptual experience.

By tracing how mechanical interaction becomes perceptual meaning, this chapter establishes the structure of the first foundational principle for haptic system design. Section 2.2 situates this foundation within the broader scope of haptic science, outlining how sensitivity and perception research together define the boundaries of tactile knowledge. Sections 2.3 and 2.4 trace the biological and biomechanical flow of tactile information–from the skin's deformation through neural encoding to perceptual inference, revealing how physical and informational processes converge. Section 2.5 examines the scientific and engineering challenge of quantifying deformation and links it to the practical limits of measurement and modeling. Section 2.6 connects these foundations to psychophysics by mapping how four primary tactile dimensions—compliance, roughness, friction, and shape—emerge from mechanical cues, forming the bridge between skin mechanics and perceptual representation. Finally, before concluding in Sect. 2.8, Sect. 2.7 highlights the multimodal and integrative nature of haptics.

2.2 Scope and Framework of Haptic Science

Haptics seeks to understand how humans detect, transmit, and interpret physical contact with the environment. Over the past few decades, advances in neuroscience, biomechanics, psychophysics, and engineering have contributed insights into the underlying principles of how humans detect and process tactile information to form perception—an adaptive process shaped by both bottom-up sensory input and top-down cognitive interpretation.

Research in haptic science is often divided into measuring tactile sensitivity and deciphering tactile perception. While these areas are closely related, each addresses different aspects of our sense of touch and requires distinct approaches. Sensitivity research defines the thresholds and boundary conditions of the sensory system—limitations of force, distance, or orientation at which stimuli can be distinguished—using psychophysical methods like the two-point discrimination test or neurophysiological techniques such as microneurography. These studies establish the baseline capabilities of the somatosensory system. Yet, thresholds alone do not explain how humans interpret stimuli as perceptual constructs like softness, slipperiness, or roughness. That deeper understanding comes from tactile perception: the study of how brain integrates and interprets sensory signals to construct meaningful, complex experiences.

2.2.1 Tactile Sensitivity

Tactile sensitivity research aims to assess the thresholds at which humans can detect or differentiate stimuli and is commonly evaluated by measuring the minimum force, distance, or orientation difference humans can perceive. One conventional method involves discrimination tests, such as the two-point spatial discrimination test [5, 93, 112], which measures the minimum distance at which a person can distinguish two separate points of contact, providing a measure of spatial acuity which varies across different regions of the body, or the two-point orientation test [13, 102, 105], which assesses the minimum angular orientation difference that a person can detect, evaluating the tactile system's ability to discern subtle directional changes in force or pressure. These tests provide quantitative estimates of sensory thresholds and establish the limits of what the human somatosensory system can detect under controlled conditions.

Beyond behavioral experiments, more direct neurophysiological techniques—for example, microneurography or other electrode-based recordings placed near peripheral nerves—allow researchers to monitor the activity of individual mechanoreceptors or groups of mechanoreceptors in response to tactile stimuli [2, 76, 108]. Although these methods can produce highly precise data, they are inherently invasive, complex, and time consuming [95, 104]. Together, these studies establish the baseline capabilities of the human haptic system—the physical limits within which perception operates.

2.2.2 Tactile Perception

While tactile sensitivity defines the limits of what humans can detect, tactile perception extends beyond these thresholds to explore how humans interpret and make sense of complex physical interactions. It is not only a question of whether a stimulus is detected, but also what it feels like. This process involves sophisticated signal processing, where multiple receptor types, neural pathways, and cognitive mechanisms converge to construct rich, multidimensional perceptual experiences, transforming mechanical events into discrete perceptual outcomes such as "soft," "rough," "slippery," or "hot".

The somatosensory system is responsible for encoding and interpreting a broad range of sensations generated by mechanical, thermal, and chemical interactions with the body. These sensations can be broadly categorized into three sensing modalities:

- Discriminative Touch: Involves sensations such as touch, pressure, and vibration, which are closely tied to the mechanical deformation of skin surface during contact.
- Proprioception: Refers to the internal sense of body position and movement, this modality arises from receptors embedded in muscles, tendons, and joints that respond to stretch and tension, informing kinesthetic awareness.
- Pain and Temperature: Includes the perception of pain intensity and temperature severity, as well as sensations like itching and tickling, detected by nociceptors and thermoreceptors. This modality encodes information about potentially harmful mechanical or thermal stimuli, enabling protective responses and thermoregulation.

Each of these categories relies on specialized receptors and dedicated neural pathways that respond to distinct physical cues. Yet they all share a common origin—the mechanical interaction between the skin and the external world. While the study of sensitivity and perception defines the boundaries of tactile experience, understanding how these signals flow through the biological system requires examining the pathways that link mechanical stimulation to neural and perceptual processing.

2.3 The Flow of Tactile Information From Skin to Brain

When the skin comes into contact with an object, mechanical forces produce localized deformation patterns of biological tissues, both at the skin surface or within deeper musculoskeletal structures, that reflect the object's physical properties, such as compliance or texture. This deformation is sensed by populations of specialized receptors known as mechanoreceptors, each tuned to respond to particular deformation patterns such as indentation depth, vibration frequency, or stretch direction.

These receptors transduce mechanical cues into neural signals, which are conveyed by afferent fibers to the spinal cord and ascend through central somatosensory pathways toward higher cortical processing centers, including the thalamus and primary somatosensory cortex (S1) [1, 83]. Along this trajectory, signals are filtered, weighted, and integrated depending on context, attention, and adaptation. The resulting neural representation is further integrated with information from other sensory systems such as motor, vision, and audition, to construct a coherent percept of the tactile event.

The process begins at the skin, which functions as a dynamic, multimodal sensing interface. The skin is embedded with distinct classes of low-threshold mechanoreceptors—such as such as SA-I (Merkel cells), RA-I (Meissner corpuscles), SA-II (Ruffini endings), and RA-II (Pacinian corpuscles)—each with unique receptive field sizes, temporal responses, and tuning to particular deformation patterns. When the fingertip contacts an object, localized deformation patterns encode physical properties such as compliance or texture. These patterns activate overlapping populations of mechanoreceptors, which generate distributed neural signals that reflect both spatial and temporal characteristics of the tactile event.

Importantly, these signals do not travel passively to the brain. Instead, they are subject to hierarchical modulation: inhibitory and excitatory processes at the spinal and thalamic levels reshape the incoming data stream, amplifying salient features and attenuating noise [24]. As the signals reach cortical areas, they are further synthesized into perceptual categories, influenced by both bottom-up sensory data and top-down expectations or learned priors [25, 75]. Thus, the brain does not merely decode touch—it actively constructs it through a process of inference, drawing upon both mechanical inputs and contextual knowledge to form coherent perceptual representations.

Understanding this flow of tactile information—from biomechanical stimulation to perceptual interpretation—highlights how the physical and neural principles of touch jointly shape perception. It reveals the necessity for haptic systems to go beyond raw data acquisition and incorporate perceptually meaningful encoding strategies, inspired by neural processing. Touch is not a simple transfer of data but an active, hierarchical process. To model and reproduce it, we must therefore examine the biomechanical foundations that determine how tactile information is encoded in the first place.

2.4 The Biomechanical Basis of Perception

Skin mechanics offers a uniquely comprehensive lens for studying tactile perception by linking physical interactions with perceptual experiences. Because people interact with objects and devices primarily through their skin, the focus on skin-level deformation provides a rich and high-fidelity source of information, capturing the intricate details of touch in a way that is non-invasive yet deeply correlated with psychophysical outcomes. This focus naturally aligns with how people experience everyday tactile interactions—grasping, pressing, and dragging motions that involve continuous contact and deformation of the skin's surface. By studying these interactions through the lens of skin mechanics, we can reveal fundamental insights about how tactile stimuli are translated and encoded into perceptual experiences.

While the flow of tactile information describes how sensory signals are transmitted from the skin to the brain, the critical insight lies in understanding how those signals originate. The mechanics of the skin form the foundation of tactile perception. How the skin deforms under contact determines which mechanoreceptors are activated, how neural signals are encoded, and ultimately how tactile experience is formed.

Unlike vision or audition, where sensory organs (the retina, cochlea) provide relatively uniform transduction, touch is mediated through a distributed, heterogeneous, and mechanically complex organ. Human skin is nonlinear, viscoelastic, and spatially varied, meaning that identical physical stimuli can evoke very different patterns of deformation depending on location, applied force, tissue composition, hydration, or even individual differences. These deformation patterns constitute the "raw data" of touch, serving as the mechanical substrate from which perceptual dimensions such as compliance, texture, and shape emerge.

To decode the sense of touch, we must first decode the mechanics of the skin. By focusing on skin mechanics, we can identify which deformation signatures correspond to specific perceptual experiences. This approach transforms touch into a quantifiable, model-based process rather than a subjective, trial-and-error exercise. It establishes a first-principle connection between biomechanics and psychophysics, enabling researchers and engineers to understand how physical input maps to perceptual output—a foundation for replicating human-level sense of touch in artificial haptic systems. Yet even with a clear conceptual framework, a major obstacle remains: measuring and modeling the physical deformation of skin with sufficient fidelity to link it reliably to perception. This challenge defines the next frontier of haptic science.

2.5 Quantifying Deformation: A Central Challenge in Haptics

While skin mechanics offer a compelling lens for understanding tactile perception, a significant challenge remains: quantifying the deformation of skin in ways that meaningfully link mechanical input to perceptual outcomes. The dynamic mechanical deformation of the skin (through stretching, compressing, and shearing) forms the physical basis of what we perceive during tactile interactions. These mechanical cues directly modulate the activity of embedded mechanoreceptors, which encode neural signals that are relayed to and interpreted by the central nervous system [82, 88] to produce rich perceptual experiences such as the compliance of an object, the roughness of a texture, or the detection of a slipping surface [15, 50, 103].

By quantifying how the skin deforms during tactile interactions and linking these deformations to neural and perceptual responses, researchers can gain critical insights into the mechanisms of touch. However, characterizing these deformation patterns remains a profound scientific and engineering challenge. Human skin is a structurally complex and mechanically nonlinear material. It is heterogeneous across anatomical regions, anisotropic in response to direction-dependent forces, and exhibits viscoelastic behaviors that vary with age, hydration, and tissue composition. These properties make the behavior of the skin under load difficult to model, predict, or measure precisely. As a result, developing quantitative, reproducible metrics that capture skin deformation with sufficient temporal and spatial fidelity remains a central scientific and engineering challenge in haptics.

This inherent complexity of skin mechanics is one of the primary reasons why the field of haptic science has progressed more slowly compared to vision and audition. Unlike the visual and auditory systems which benefit from specialized, uniform sensory organs like the retina and cochlea that exhibit relatively well-defined and consistent transduction mappings, the somatosensory system is spatially distributed and context-dependent. The somatosensory system integrates signals from multiple classes of mechanoreceptors, each located at different depths and densities within the skin. These receptors are embedded in layers of tissue that vary significantly in stiffness, thickness, and elasticity across individuals and even across a single fingertip [30, 36, 80]. As a result, two users interacting with the same object under similar conditions may produce different skin deformation profiles and consequently different neural responses and perceptual experiences. This variability—rooted in the complex interplay of biomechanical and neurophysiological factors—has historically limited our ability to build predictive models of touch and hindered the development of generalizable high-fidelity haptic technologies.

Addressing these challenges is essential for the progress of haptics. It requires not only technical advances in sensing and actuation but also deeper foundational knowledge of how mechanical inputs are transduced into meaningful perceptual states. In other words, we must understand not just how the skin deforms, but which deformation features carry perceptual significance, and how those features are encoded by the nervous system. The key lies in identifying the deformation cues that correlate with distinct tactile dimensions—importantly, those that describe how humans access an object's compliance, roughness, friction, and shape.

Having established why quantifying deformation is central to advancing haptic science, we now turn to the perceptual dimensions through which touch expresses meaning. The considered dimensions—compliance, roughness, friction, and shape—provide measurable links between biomechanical cues and psychophysical experience. By analyzing the scientific literature on deformation cues and their neural encoding, we will draw connections between fundamental haptic science and practical engineering design goals. These insights will inform how future haptic systems, particularly those grounded in first principles, can be built to emulate and respond to the real perceptual structure of human touch.

2.6 Linking Skin Biomechanics to Psychophysics

Human touch enables the assessment of an object's mechanical, textural, and geometric properties, supporting actions from stable grasping to delicate exploration. In these interactions, the perceptual ability to discriminate compliance, roughness, friction, and shape is fundamental for object classification, manipulation, and evaluation. These four tactile dimensions are chosen because they collectively capture the core functional space of human touch. Each corresponds to a distinct class of mechanical interaction between the skin and external stimuli, and together they span the major modes of contact that govern real-world tactile events. They encode how an object feels (its softness or slipperiness) and what it is (its shape or surface structure), forming a multi-dimensional perceptual space through which humans interpret the world via touch. These dimensions are abstractions for biological encoding by mechanoreceptors whose form and function are well-studied through behavioral experiments and recordings of the firing activities of these afferents [8, 57, 60, 97].

Each dimension maps to a specific set of mechanical cues—patterns of skin deformation, vibration, or force distribution—that can be quantified, modeled, and replicated via hardware, software, and application subsystems. For example, compliance perception informs sensing and actuation thresholds for force and deformation; roughness perception drives the need for high-bandwidth vibration transduction and rendering; friction guides real-time control strategies for grip and motion stability; and shape imposes demands on distributed sensing and form reconstruction.

By articulating these relationships, this section establishes a scientific foundation for defining subsystem-level design goals. These four tactile dimensions link the mechanics of touch with the neural coding of perception, enabling the development of integrated haptic systems that are grounded in how humans sense, interpret, and respond to physical contact.

2.6.1 Compliance: Perception of Hardness

Compliance perception illustrates how cutaneous deformation translates directly into perceived material properties, offering a model for how physical and neural cues merge.

Compliance perception is dominated by mechanical cues derived from skin deformation under load. It depends on how much skin surface is deformed—how deeply the finger pad indents and how largely or rapidly it extends vertically and laterally. These deformation-based cues provide critical information about how the skin interacts with objects of varying compliance, informing the nervous system to interpret material perception. Particularly, for softer objects, the finger pad deforms more extensively; for stiffer objects, the deformation is minimal and occurs quickly. While force-related cues (e.g., force-displacement ratio, terminal force, force rate) also contribute to compliance discrimination, especially for tapping motion or pressing on stiff surfaces, the majority of information appears to come from the skin's cutaneous deformation patterns themselves.

Compliance is a fundamental property of objects that humans sense and interpret through touch. Its perception involves an interplay among cutaneous deformation cues, force cues, and proprioceptive (joint and muscle) feedback, as well as the exploratory strategies adopted when interacting with an object. The study of com-

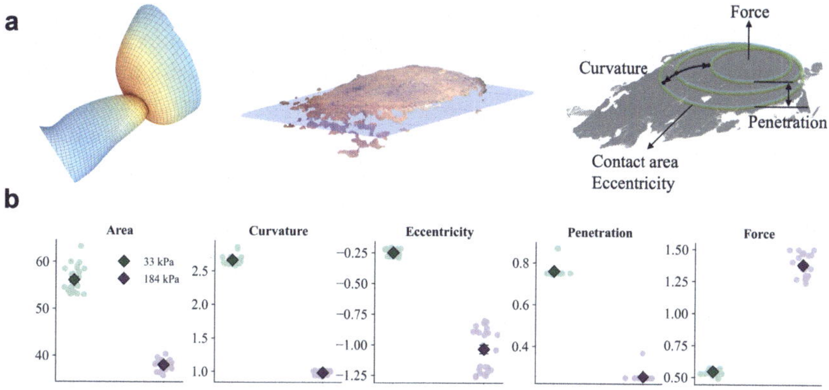

Fig. 2.1 Fingertip deformation varies with stimulus compliance. These deformation cues can be extracted using techniques including Finite Element Modeling (FEM) and high-resolution imaging systems, illustrated in (**a**). In (**b**), empirical data shows that "softer materials... [72]". Units are in millimeters (mm) for Area, Curvature, Eccentricity, and Penetration, and newtons (N) for Force

pliance perception relies heavily on advanced experimental tools and computational models to characterize how the skin responds to deformable materials. Recent research has examined how these deformation cues are encoded by mechanoreceptors in the skin, how tactile information is integrated in different contact conditions (active vs. passive, bare-finger vs. tool-based), and how perception of compliance can be influenced by factors such as velocity and contact duration.

A wide range of skin deformation cues contribute to the perception of compliance by encoding the relationship between mechanical changes on the skin surface and the compliance of the object contacted, see Fig. 2.1. These identified cues help characterize complex skin deformation patterns and show correlations with compliance perception by how they are encoded by different classes of mechanoreceptors within the skin. Psychophysical studies show that removing direct surface-deformation information (e.g., using a rigid cover) significantly reduces compliance discrimination performance, highlighting the critical role of deformation-related cues in compliance perception [65, 98].

Contact Area (Stress) As an object contacts with the skin, either in passive or active manner, the contact region grows outwards in a circumferential way. The area of contact contains powerful information for compliance perception because it directly reflects how the skin and the object share the compressive load. The expanding region of contact redistributes contact force (i.e., stress) within the skin and subcutaneous tissue, activating neuron receptors correspondingly.

Under load, SA-I afferents, with small receptive fields (RFs), track local pressure changes around the contact zone as it expands. During sustained contact, as the contact area gets larger, more SA-I units become active and neuron firing rate increases due to growing fields in their RFs. In contrast, RA afferents will respond to dynamic increase in contact area especially at the onset of contact (transient response) and they adapt if the contact area remains constant. Experimental paradigms that measure contact area via imaging or sensor techniques and conduct psychophysical evaluation of the participants have demonstrated a high statistical correlation between contact area expansion and perception of compliance [19, 71]. Overall, contact area is more than just a geometric measure. It is a driver of the stress distribution within the skin deformation patterns that directly shape the spatial firing rates of mechanoreceptors. As these afferents encode differences in local pressure and stress, the central nervous system integrates their signals to form compliance perception.

Skin Stretch (Strain) Skin stretch refers to the lateral (tangential) deformation of the skin surface that occurs when pressing on or sliding across an object. In the context of compliance assessment, skin stretch arises naturally because real interaction rarely involves perfectly normal (vertical) compression, as friction and small variations in contact angle often create shear forces. The extent and pattern of this lateral deformation depend on the object's compliance, as softer objects allow for deeper normal indentation, which also can induce larger tangential distortions in the skin. As the skin presses into an object's surface, part of the skin sinks into the object (normal deformation), while surrounding skin regions are pulled or pushed sideways (lateral or shear deformation), causing an effect of multidirectional strain distribution [71]. These strain patterns are mapped into the skin's epidermis and subdermal tissues, firing the embedded neuron afferents, and ultimately contributing to the formation of compliance perception [28, 90]

SA-I afferents, which have relatively small RFs, respond to localized gradients of deformation including shear stress within their territory. As the skin stretches, SA-I units in those active zones register changes in strain magnitude and potentially increase their firing rate. SA-II afferents, found more abundantly in the palm and less in the fingertip, have larger RFs and are highly sensitive to skin stretch and tension that span broader areas, hence detecting the overall strain patterns and capturing global lateral stretch. RA afferents can fire brief bursts upon changes in stretch rate or direction. When the skin slides or initially indents an object, a transient stretch can cause short-period RA firing, but they do not maintain sustained firing if the stretch is static.

By encoding lateral strain or tension patterns, skin stretch provides the central nervous system with an additional dimension of information about how the material supports, deflects, or resists skin surface movements. Skin stretch activates a combination of small-field (SA-I) and potentially large-field (SA-II) afferent responses, providing the brain with continuous and spatially integrated information about how the skin is deformed, which helps define the perceptual assessment of an object's compliance, especially in natural touch contexts where friction and off-normal forces are present.

Indentation Depth Indentation depth describes how far the skin sinks into an object under a given applied load. Specifically, when the skin makes contact with an object, the surface material deflects to some extent (more significantly for a soft material and less for a stiff one), allowing the skin to indent and conform to the surface. Because this indentation involves both normal and tangential deformations in the skin, the skin's layers and embedded mechanoreceptors experience increasing strain as the indentation depth grows. When indentation depth is measured directly (e.g., via 3D imaging or customized devices), it shows correlations with participants' subjective compliance discrimination [19, 20, 27, 71].

SA-I afferents fire in proportion to sustained or slowly varying skin indentation and tension within their RFs. As indentation depth increases, local compression and stretch intensity, elevating the SA-I afferents' firing rate (or recruiting additional SA-I units). This rise in population-level activity of SA-I afferents can encode how deeply the skin is depressed. If SA-II afferents are present in the scenario, a deeper indentation may create large-scale strain, prompting these afferents to increase their discharge accordingly as they respond to more global skin deformations. SA-II responses might integrate information over a wider region, helping the brain sense the overall degree of yielding to touch. RA afferents are most sensitive to the onset and rate of change during indentation, especially for tapping or short pressing. When the skin initially pushes into a surface, RA afferents produce a burst activity (e.g., neural spikes) that correlates with how fast the indentation depth is increasing. Once the indentation depth reaches a plateau, RA firing diminishes, making these neurons less informative about the sustained depth but useful for detecting fast changes.

Indentation depth provides a direct mechanical link to how much the contact surface yields to touch. Mechanoreceptors (particularly SA-I afferents) transduce greater local strain into higher firing rates or expanded recruitment, enabling the central nervous system to interpret signals related to compliance.

Surface Curvature When the skin presses against an object, the local curvature of the skin surface—how sharply or smoothly the skin bends or "bulges" under compression—can vary dramatically depending on the object's compliance. A high-curvature region indicates a relatively sharp bend, whereas lower curvature is more gradual. Curvature differences are particularly pronounced in the soft-to-medium compliance range, where the skin undergoes enough deformation to reveal spatial nuances in skin deformation patterns. This spatial pattern of curved or flattened skin regions can be measured using 3D imaging techniques (e.g., stereo-camera setups) and studies show that surface curvature can predict how participants rate or discriminate the compliance of objects [40, 71].

Mechanoreceptors (especially slowly adapting types) respond well to localized surface curvature in the skin, and their firing rates can encode the magnitude of that curvature. Sharp changes in curvature produce heightened compression or tension within that region, during increased neural firing of SA-I units. When the skin is more gradually deformed (typical of softer objects), many SA-I units across a broader area are moderately activated. Harder objects can create a more localized "peak" in curvature, producing highly focal SA-I activation patterns. RA afferents can register curvature changes especially during dynamic transitions (e.g., tapping or onset and offset of contact), but then adapt as the contact stabilizes in the sense that it does continue to fire in response to sustained stimuli. For continuous contact interaction, SA-I units are believed to be the main channel for curvature-related steady-state signals.

Changes in local surface curvature offer a fine-grained spatial signal about how the skin and object are interacting. Leveraging 3D imaging and modeling techniques, research has shown that skin surface curvature correlates well with compliance judgments, especially for softer materials [71, 72]. From a neurophysiological standpoint, SA-I afferents capture these subtle gradients in skin strain and their combined population response is crucial for encoding how deeply the skin surface is bending. Ultimately, this curvature-elicited neural coding helps the brain parse whether an object yields gently (soft) or resist sharply (stiff), thus providing one of the primary building blocks of our tactile perception of compliance.

Contact Shape Beyond area and depth, the overall shape of the contact region (circular vs. elliptical, uniform flattening vs. localized bulges) reflects how the skin and object surfaces conform to each other. Empirical studies using 2D photographs and 3D imaging have shown a strong correlation between contact shape and perceptual sense of compliance, namely that softer materials lead to more elliptical contact shape, whereas stiffer materials tend to constrain the contact region into a more circular shape [70, 81]. Because softer objects permit deeper and more unevenly distributed deformation, the skin on the finger pad tends to spread out more along its long axis (parallel to the finger's length), forming an elliptical shape on the contact surface. The extra freedom to "sink in" tangentially also fosters greater lateral skin stretch, reinforcing an elongated contact region. In contrast, stiffer materials limit overall indentation, confining the skin contact to a more symmetrical, circular region. Since the deformation is shallower and the skin cannot spread extensively, the contact shape often stays compact and nearly round.

SA-I afferents fire in response to local indentation and skin stretch within their RFs. A more elliptical contact region means that certain parts of the fingertip periphery experience additional shear or stretch, activating SA-I units across that elongated boundary. In contrast, a circular shape confines intense strain to a smaller or more centralized activation zone. While fewer in number at the finger pad, SA-II afferents are sensitive to larger-scale tension patterns. They may help distinguish between elongated vs round shapes if there is enough broad skin stretch. RA afferents respond strongly during changes in contact shape, especially if the finger taps or shifts on the surface, or pressing into a less compliant object. However, once the change of contact shape is stable, RA firing diminishes. Thus, they can help detect how quickly the shape transition occurs which underlies the perceptual discriminability in compliance.

Contact shape offers a spatial snapshot of how the skin (at the finger pad) and object interact under load. Softer objects often form an elliptical shape that mechanoreceptors interpret (via changes in local strain and surface stretch) as indicative of greater compliance. By contrast, a circular contact shape encountered by stiffer objects signals a more symmetric, concentrated deformation region. These shape-related differences in skin deformation patterns become encoded in distinct neuron spiking activities. As such, contact shape is another important dimension of skin deformation cues that helps the central nervous system differentiate variations in compliance.

Contact Force Traditionally, contact force is thought as a straightforward, external variable: the load measured at the surface between skin and object. However, the perceived force in the context of compliance perception is fundamentally tied to how the skin itself deforms, distributes stress, and ultimately activates mechanoreceptors. In this sense, contact force can be viewed as a macroscopic reflection of underlying skin deformation.

As the finger pad interacts with an object (via pressing or tapping), contact force arises when the skin is indented, generating a stress distribution that is integrated across thousands of mechanoreceptor units. SA-I afferents track local compression or tension, and RA afferents respond to changes in that load. Collectively, their activity communicates a sense of "force magnitude" to the brain. For stiffer objects, even a minor indentation ramps up force quickly, producing intense local stress and strain in a small area. Mechanoreceptor firing patterns (especially RA afferents detecting rapid on- and off-sets, and SA-I afferents sensing sustained, localized pressure) are interpreted as a combined, strong "force" signal, leading to the observer to perceive and differentiate between stiff materials. On contrary, when interacting with soft materials, the skin deforms more, distributing stress over a broad area. Hence, force differences can remain small until larger displacements occur. This reduced force signal leads the observers to reply on cutaneous cues such as contact area or skin surface curvature.

Indeed, empirical studies have demonstrated that in passive touch, contact force, as a skin deformation cue, contributes significantly to differentiating the elastic substrates that are stiffer than the skin; while discriminating between stimuli that are softer than the skin, contact force becomes less informative as the participants shift to relying on surface-deformation cues[39, 71, 72]. Thus, contact force is not merely an external load measurement; it is deeply intertwined with the mechanical deformation of the skin. The brain infers "force level" from the spatial and temporal patterns of receptor firing, which are byproducts of how the skin has deformed. For stiffer materials, force emerges as a prominent channel (a macro-level skin deformation cue) that helps participants distinguish compliance differences, whereas for softer objects, the brain leans more on direct indicators of skin deformation to gauge compliance variations.

The perception of an object's compliance is derived from the integration of several key cutaneous cues generated by skin deformation under load.

- **Contact Area:** Softer objects create a larger, often elliptical, contact area, increasing activation of SA-I afferents.
- **Skin Stretch:** Lateral deformation of the skin provides strain information, encoded by SA-I (local) and SA-II (global) afferents.

(continued)

- **Indentation Depth:** How far the skin sinks into an object is directly transduced by SA-I afferents, with deeper indentation indicating softer materials.
- **Surface Curvature:** The sharpness of the skin's bend varies with compliance, creating distinct spatial patterns encoded by SA-I afferents.
- **Contact Shape:** Softer materials lead to more elliptical contact shapes, while stiffer ones produce more circular shapes, altering strain patterns.
- **Contact Force:** For stiff objects, force is a dominant cue. For soft objects, its role diminishes in favor of the direct deformation cues above.

Ultimately, the brain integrates these spatially and temporally distributed neural signals to form a coherent perception of compliance.

2.6.1.1 Deformation-Driven Neuron Coding: Independent and Complementary Contribution to Compliance Perception

Multiple skin deformation cues, such as contact area, curvature, penetration depth, and force, activate diverse populations of mechanoreceptors, each tuned to specific aspects of physical stimulation. Slowly adapting type afferents, for instance, are sensitive to sustained pressure and spatial variations in deformation, effectively capturing the magnitude of contact area or how deeply the skin penetrates into the material. In contrast, rapidly adapting afferents respond robustly to dynamic or vibratory stimuli, such as abrupt changes in force. By parsing information in parallel, these receptor types function as distinct yet complementary channels of input: each cue, be it spatial or temporal, sustained or transient, provides an independent source of data on perception of compliance.

Despite this independence, these cues also reinforce one another to enrich compliance perception. An SA-I afferent, for example, may escalate its firing rate when local skin curvature intensifies, indicating increased pressure on a particular skin region, while an RA afferent in the same region may spike in response to a rapid expansion of contact area. Meanwhile, SA-II or deep mechanoreceptors (e.g., Pacinian corpuscle units) can detect broader strain patterns and high-frequency vibrations, further refining the underlying percept. Through this division of labor, the sensory system distributes computational demands among specialized pathways—SA-I units focus on spatial details and local gradient changes, RA units on transients and rate signals, and SA-II afferents on global strain variation–allowing the central nervous system to integrate multiple streams of relevant data without overloading any single channel.

The result is a perceptual synergy: each cue has the capacity to independently encode facets of compliance, yet together they form the nuanced, context-sensitive percept of softness or hardness we rely on in everyday touch. As contact conditions vary, such as changes in exploration strategy, velocity, or the material's modulus, the brain can re-weight these cues dynamically, adjusting the relative emphasis on static versus transient signals. In this way, the independent roles of each cue remain a cornerstone of tactile coding, but it is ultimately their complementary interplay that furnishes the richly detailed perception compliance.

Consequently, to build effective haptic devices that replicate such intricate perception of compliance, one key design principle is to offer multiple channels of feedback. Instead of a single force-displacement loop, systems can simultaneously provide information such as localized penetration depth, dynamic contact area change rate, or transient vibrations. Such parallel processing mirrors the division of labor in the nervous system, allowing each cue (spatial and temporal) to stand alone while also reinforcing the realism of the overall tactile experience. By splitting the management of these cues among different actuation mechanisms or control loops, the system can avoid overloading any single signal path.

Another important implication for design arises from how the haptic system re-weights these cues depending on specific physical interactions. To reproduce different levels of compliance, softer objects may require large displacements and expansive contact area simulations, whereas stiffer objects can rely on a sharper change on force and contact shape. Designers can incorporate hardware and software subsystems that dynamically adjust these channels accordingly to real-time user input and feedback, thus aligning with the brain's strategy of selectively emphasizing certain cues on a contextual basis.

A further consideration is the benefit of using fine-grained or distributed actuation mechanisms. Because of the high spatial distribution of mechanoreceptors that are integral to cutaneous interpretation of compliance, a single-point actuator or uniform surface modulation may fail to capture the details the nervous system expects. Instead, allowing multiple, independently controlled actuation components can create defined differential shear or indentation across the skin contact region. This local variance better approximates the naturally occurring patterns of skin deformation that the corresponding mechanoreceptors can encode.

2.6.1.2 Spatial and Temporal Aspects of Compliance Perception

Our skin is viscoelastic (i.e., time-dependent in its stress-strain response) and nonlinear (its stiffness changes with deformation). As a result, when the finger pad presses onto an object, even if force is held constant, the skin continues to deform or "creep", altering the stress distribution and skin deformation [23, 53]. Additionally, a fast indentation produces a different strain pattern, and thus different neural activation than the same final indentation reached more slowly [72]. Because of these properties, compliance perception cannot be captured by a single, static measurement (e.g., just the final force or the terminal contact area). Instead, the participants must integrate how the skin and object together evolve over time.

Neural Coding: Spatiotemporal Patterns of Firing Correspondingly, cutaneous mechanoreceptors (e.g., SA-I, RA) encode mechanical events in their neuron firing trains. Neuron firing trains are the specific temporal sequences of electrical impulses, known as action potentials or spikes, generated by a neuron over a period of time and can be described by [56]:

- Firing Rate (Temporal). The firing rate is the instantaneous or average firing frequency over time. For instance, a steep contact force increase can produce a high initial firing rate in RA afferents.
- Number of Spikes (Spatial Summation). Over an entire contact period, the total spike count (for an individual neuron or a group) can reflect the degree of deformation. Soft materials, which allow deeper indentation, may cause higher total spikes across SA-I units.
- Firing Duration (Temporal Window). Firing duration reflects how long a neuron remains active at elevated firing rates. Materials that sustain ongoing deformation cause prolonged firing in slowly adapting afferents (SA-I, SA-IIs).
- Spike Latency (Temporal Onset). Spike latency describes the delay between stimulus application and the first spike(s). For example, a stiff object can cause a short latency due to rapid high stress accumulation, while a gradual indentation into a soft object might delay neuronal responses.
- First Firing Time and Peak Firing Amplitude (Temporal Dynamics). This parameter shows when the firing peaks and how strong that peak is. If the skin is indented quickly, RA afferents can produce a strong early peak in firing, reflecting the rate of force buildup.

In practice, the brain receives a constellation of these spatiotemporal signals from a population of neurons. Hence, compliance perception emerges from both the magnitude of deformation (spatial) and the dynamic evolution of mechanical cues over time (temporal)

Terminal vs. Time-varying Cues Many experiments investigating perception of compliance have focused on end-point measurements, termed terminal-phase measurements, that can be discrete or static. Participants typically press their finger (or by an indenter) to a certain force or displacement and then measure final contact area or indentation depth. For example, customized elastic slabs are fabricated with varying microstructure (e.g., thickness, height) and used to measure contact area and indentation depth under specific loads or displacements. Similarly, soft foam and pressure sensor films have been used to record contact force and contact area after each interaction with the object [19, 22]. These approaches illustrate correlations between deformation mechanical cues (e.g., contact area, indentation depth) and perceived compliance. However, the discretization, often just a handful of data points at different loads or time interval, is too coarse compared to the continuous and instant changes the skin undergoes. Human can sense differences in force or spatial variation within tens of milliseconds, yet many static paradigms can effectively take measurements only at large intervals. As a result, temporal dynamics, e.g., rate of contact area change or force build-up, are not captured. This limits the fidelity and accuracy with which such paradigms can reveal why or how compliance is perceived.

By contrast, some recent paradigms use high-speed cameras and 3D imaging at sufficient temporal (e.g., >10–30 Hertz) and spatial (e.g., 60–120 micrometer resolution) rates to record real-time surface changes in the skin [40, 71]. These methods track ongoing evolution of mechanical cues derived from skin deformation during contact events, closely matching the continuous, dynamic signals that human mechanoreceptors receive. By combining spatial (magnitude) and temporal (rate, onset, offset) dimensions, the perceptual system arrives at a nuanced sense of compliance, revealing how "fast", "deep", and "wide" the skin is deforming. Indeed, these time varying, spatiotemporal skin deformation cues—such as the rate of contact area, skin surface curvature, and indentation depth—have emerged as more powerful predictors of perceived compliance than the end-point, terminal measurements in psychophysical and neurophysiological studies [39, 71, 72, 113].

In summary, the perception of compliance is not a simple matter of measuring force, but a sophisticated neural interpretation of complex, dynamic skin deformation. When we press into an object, the viscoelastic skin deforms in a multi-dimensional pattern, generating key mechanical cues—contact area, skin stretch, indentation depth, surface curvature, contact shape, and contact force—that are directly encoded by populations of specialized mechanoreceptors. SA-I afferents provide a high-fidelity spatial map of sustained pressure and local strain gradients; RA afferents capture the critical transient dynamics and rates of change; and SA-II afferents contribute information on larger-scale skin tension. Crucially, the perceptual system relies more heavily on the rich, continuous stream of time-varying cues (like the rate of area expansion or force build-up) than on static, terminal-phase measurements, as these dynamics more closely match the real-time signals processed by the mechanoreceptors. The brain continuously integrates these parallel spatiotemporal signals to form a robust percept of the compliance of an object.

2.6.2 Roughness: Perception of Surface Texture

Similar to compliance, roughness perception arises from how skin deformation and vibration encode texture. Roughness perception plays a fundamental role in our ability to recognize and interact with textured surfaces. From the texture of fabric to sandpaper, roughness enables humans to discern subtle variations in material composition and structure through touch. This ability is essential not only for basic sensory exploration, but also for complex tasks requiring precision and adaptability, such as tool manipulation, surface inspection, and object recognition.

Scientifically, the study of roughness offers profound insights into how the human somatosensory system, comprising mechanoreceptors in the skin and higher cortical processing centers, interprets and categorizes the microscopic (fine) and macroscopic (coarse) features of surfaces. This knowledge helps unravel how spatial and temporal patterns of neural activity translate into perceptual experiences and also sheds light on broader questions about how sensory systems convert physical stimuli into perceptual outcomes, contributing to our understanding of human sensory integration and neural coding. From an engineering perspective, roughness is a fundamental consideration in the design of haptic technologies. In virtual and augmented reality systems, accurate reproduction of roughness is critical for creating immersive and realistic tactile experiences. In robotics, the ability to detect fine textural features and replicate roughness sensation enhances the dexterity and sensitivity of robotic manipulators, enabling them to perform tasks ranging from industrial assembly to medical procedures. There are a number of key facets in understanding roughness as it relates to tactile perception.

Vibration When the finger (or probe) moves across a textured surface, microscopic ridges induce vibrations in the skin. These vibrations are critical for finer features and are preferentially encoded by RA afferents and Pacinian corpuscles. Studies of adaptation confirm that when vibrotactile pathways are fatigued, roughness perception of fine textures is diminished [47, 49].

Contact Area For bare-finger interactions, the deformed skin area in contact with the surface contributes to spatial coding. The overall contact area depends on the skin's physical states, contact force, and local friction. Larger surface contact can dampen high-frequency vibrations and alter how many mechanoreceptors are stimulated. With probes, the size and shape of the probe tip determine the effective contact area and influence roughness perception. A smaller tip can fall between raised elements more readily, causing discrete, high-amplitude vibrations. In contrast, larger or flat probes may mimic more of a "bare-finger pattern", dispersing contact over multiple surface features.

Contact Force (Normal and Tangential)

- Normal Force: Enhances spatial coding by increasing skin deformation and affects skin indentation depth. Higher force typically intensifies roughness perception up to a threshold, likely by increasing local deformation of the skin [12, 68, 119].
- Tangential Force: Oscillations in tangential force can closely track surface ranges and influence temporal coding by modulating the amplitude of vibratory signals during lateral movement. Studies have shown that the rate of change in tangential force strongly correlates with subjective roughness judgments, highlighting the importance of shear-related cues [63, 96].

Skin Displacement The magnitude and pattern of skin displacement encode critical information about surface features: specifically, how much the skin is deflected and in what pattern by the surface during contact. Larger groove widths and deeper corrugations can produce stronger, more localized deformations. This leads to enhanced differences in firing across SA-I and RA afferents [37, 58, 69].

In many cases, spatial and temporal cues work in concert, depending on the texture's spatial scale to form a unified roughness percept. However, since afferent responses to spatial features are shaped in complex ways by skin mechanics and variations in exploration strategies, it is therefore difficult, if not impossible, to draw definitive conclusions about neural coding in touch without analyzing the skin tissue response under different contact conditions.

Direct vs. Indirect Touch

- Direct Touch: When the bare finger (glabrous skin) contacts the surface it generates spatial coding from the SA-I afferents. For coarse features (>200 micrometers), exploration through direct touch enables robust roughness constancy across speeds and enables better discrimination among surfaces than exploration through indirect touch [11, 62]
- Indirect Touch: Alternatively, under indirect touch, vibrations to the hand are mediated, e.g., by a cloth, probe, or rigid link. For indirect touch, Pacinian corpuscles and RA pathways dominate spatial coding. For instance, holding a rigid stylus over a textured surface can shift the perceived maximum roughness depending on the geometry of the probe tip relative to the surface features [63].

Bare Finger vs. Probe The choice between bare-finger exploration and probe exploration, a specific kind of indirect touch, has significant implications for which mechanoreceptor populations and which coding mechanism (spatial vs. temporal) are most active.

- Bare Finger: Yields a spatial map of contact. Larger contact area between the skin and the surface enhance spatial coding for coarse textures. Skin states such as compliance and geometry, contact force, and friction strongly influence the magnitude of local skin deformation.
- Probe: The tip geometry (e.g., diameter, shape) and scanning approach determine how vibrations are relayed to the skin, altering roughness perception. A smaller tip may slip between raised elements, generating a characteristic "drop point" that can amplify vibratory intensity at certain spacing [62, 63].

The perception of surface roughness is a complex synthesis of spatial deformation and vibratory signals, shaped by how we interact with a surface. Unlike compliance, which is dominated by normal indentation, roughness perception critically depends on lateral motion, which generates both a spatial strain pattern and micro-vibrations. The brain integrates these cues to form a coherent texture percept, with the relative importance of spatial and temporal coding shifting based on the texture's scale and the method of exploration. Key facets that define roughness perception include:

(continued)

- **Vibration:** Micro-vibrations induced by lateral motion are encoded by RA and PC afferents and are the primary cue for perceiving fine textures.
- **Spatial Cues:** For coarse textures, the spatial pattern of skin deformation—captured by contact area and skin displacement—is encoded by SA-I afferents to create a neural "map" of the surface.
- **Exploratory Forces:** Both normal force (influencing skin deformation) and tangential force (modulating vibration amplitude) are actively controlled by the perceiver and directly impact the intensity of roughness.
- **Mode of Touch:** The neural code for roughness changes dramatically between direct and indirect touch. Bare fingers leverage both spatial and temporal cues, while tool-based exploration (e.g., with a probe) relies more heavily on vibratory channels, filtering the information through the tool's geometry.

Ultimately, roughness is not a single property of a surface but a perceptual experience constructed from the interaction between surface geometry, skin mechanics, exploratory action, and the specialized tuning of our mechanoreceptive channels.

2.6.3 Friction: Perception of Stickiness, Slipperiness, and Dryness

Skin deformation depends heavily on the friction of the surface being touched. The perception of friction is dominated by the interaction dynamics between skin and surface. The study of friction utilizes a combination of experimental, computational, and neurophysiological methods to analyze the mechanical interactions between the skin and contacted surfaces. These methods help quantify frictional force, skin deformation patterns, neural responses, and vibratory signals, providing a comprehensive understanding of how humans perceive friction.

High-resolution imaging allows researchers to directly visualize skin deformation and contact interactions at the skin-surface interface in detail. Optical imaging systems, such as high-speed cameras, enable the tracking of 3D micro-geometrical features of the skin during contact events and examination of how these interactions influence friction perception. For example, high-speed imaging methods equipped in a controlled mechanical platform can be used to analyze how the real contact area changes during slip onset, capturing detailed friction-induced skin deformation, and to detect localized strain patterns in the skin, revealing micro-slips that precede full slip events [14, 16]. Optical coherence tomography (OCT) is another approach, allowing cross-sectional views of the skin's subsurface layers during sliding, allowing quantification of internal stress distributions associated with tangential loading [17]. First, we review a variety of methods related to the study of the tactile perception of friction.

Direct Measurement Tools To capture the frictional forces occurring during physical interactions, a variety of direct measurement tools are employed:

- Load cells and force sensors: Multi-axial load cells and force/torque sensors are frequently used to measure the normal and tangential forces that arise when a finger interacts with a surface. Studies have used force sensors to analyze how grip force adjustments correlate with perceived friction [38, 77]. These measurements can help quantify the force ratio needed to maintain grip stability, revealing how humans adapt their grip to different materials.
- Tribological instruments: Specialized instruments (e.g., tribometers) can directly measure the coefficient of friction (CoF) and adhesion forces between the skin and a material surface. Tribometers can be used to study how skin properties (e.g., hydration) and surface features influence perceived friction [79].
- Pressure films and tactile sensors: These tools can be used to capture spatial variations in force and pressure distribution across the contact region. They are particularly useful for studying anisotropic effects in skin deformation, where frictional forces vary depending on the direction of movement relative to skin microstructure [14, 18, 43].

Simulation Methods While direct measurement tools and imaging techniques provide data from real-world interactions, simulation methods allow systematically manipulating variables that might be difficult or impossible to control in practice. These computational models can provide a deeper understanding of friction perception by simulating skin mechanics, force distribution, and neural encoding.

Finite element models (FEMs) are widely used to simulate skin deformation and stress distribution under different frictional conditions. Studies have used FEMs to analyze how different orientations of subcutaneous fiber influence contact pressure distribution and partial slip detection, providing key insights into frictional signal encoding [43]. Additionally, mechanical and physical models permit replication of human skin's physical property and geometry, mimicking skin deformation patterns caused by distributed force during contact and predicting perceptual effects based on the simulated mechanoreceptors activities [28, 46].

Microneurography: Neural Recordings of Friction Perception Beyond physical and mechanical perspectives, the sensory dimension of friction is central to our sense of touch. Microneurography, a technique that directly records action potentials of nerve activity, bridges the gap between physical measurements and neural encoding. Pioneering work has been done to reveal how different mechanoreceptor units innervating at the fingertips respond to slip event [15, 74]. Through this technique, friction-related mechanical events, such as the onset of slip, can be correlated with specific firing patterns in peripheral nerves. As a result, microneurography offers a powerful tool for linking perceptual reports of slipperiness and stickiness to underlying physiological mechanisms, complementing the engineered realism of simulation methods and the quantitative precision of direct measurement tools.

Frequency Analysis of Friction-Induced Vibrations Frictional interactions are not just about forces; they also generates dynamic vibratory signals that activate mechanoreceptors. Frequency-domain analysis techniques, such as Fourier transforms and wavelet decomposition, can be used to provide valuable insight into how friction is translated into sensory input. Studies have demonstrated how micro-vibration, ranging from low to high frequency, can carry critical information about surface slipperiness, texture, and micro-slip events [26, 48, 109]. By examining the spectral content of these vibrations using accelerometers or special sensors, distinctive vibrational profiles can be characterized under different surfaces or sliding velocities.

A combination of imaging, force measurement, simulation, neurophysiological, and vibratory analysis techniques provides a comprehensive framework for studying friction perception. High-resolution imaging captures skin deformation in real time, while direct measurement tools quantify mechanical parameters such as force, shear stress, and contact area changes. Simulation tools help predict mechanical and neural responses,and microneurography links physical measurements to neural coding. Moreover, frequency analysis of friction-induced vibration uncovers the role of dynamic tactile signals in friction perception and slip detection. By integrating these methodologies, our understanding of friction perception can be comprehensively studied, setting up scientific foundation for advanced, effective haptic system design.

2.6.3.1 Mechanical Cues and Their Underlying Neural Coding that Influencing Friction Perception

The detection of friction is primarily driven by the dynamic and spatiotemporal changes in skin deformation, not by the absolute magnitude of tangential or shear forces. In other words, it is the way in which the skin's deformation pattern suddenly changes that informs the neural coding of friction.

Coefficient of Friction In precision grip tasks, the static CoF , defined as the ratio of the tangential force at slip onset to the normal force, plays a crucial role in determining the minimal grip force required to securely hold an object. Contrary to the traditional tribological theory for many solid materials that treats the CoF as a constant value (independent of the normal force owing largely to surface roughness), research on human skin—especially fingertip skin—shows that CoF varies nonlinearly with contact pressure and normal force. Specifically, studies show that the CoF increases as the normal force decreases, a behavior that can be described by a negative power law [3, 22]. It indicates that when the normal force is relative low, friction coefficients become higher, contributing to a "sticky" feeling. Conversely, at higher normal loads, the CoF tends to drop, suggesting a shift in the dominant friction mechanisms at play.

The primary explanation for this non-constant, nonlinear relationship lies in the role of adhesion. At low normal forces, the adhesive interaction between the skin and the contacting surface become more pronounced. This is particularly evident on smooth, impermeable surfaces like glass, where extended contact time duration at lower loads allows moisture (in the form of sweat) to occlude the skin's topographical features [18]. Such occlusion softens the surface ridges of the fingertip, effectively enhancing adhesion. As a result, the adhesive forces—arising from molecular interactions across a real area of contact, rather than mechanical interlocking due to roughness—dominate the frictional perception and behavior.

The relationship between the CoF and contact pressure for the human fingertips is well captured by both experimental measurements [3] and contact mechanics modeling [22]. An increased CoF at low normal forces means that individuals can achieve a secure grip with relatively light touch, which is particularly advantageous during precision tasks where excessive force might be detrimental. The nonlinear relationship between CoF and contact pressure thus allows the nervous system to fine-tune grip force dynamically, ensuring that objects are held securely while minimizing unnecessary exertion. By dynamically shifting between adhesion-dominated friction at low loads and deformation-based friction at higher loads, human fingertips are capable of performing fine grip-control tasks such as gentle grasping of fragile objects and adapting to object properties during manipulation.

Friction Force Friction force is the total force experienced at the skin-surface interface. Importantly, forces arising from micro-scale deformation of the skin and adhesion at micro-junctions can contribute significantly to this overall friction. In the context of human tactile interactions, especially those that involve the finger pad, friction force is not a fixed material property. It arises from two main components:

- Adhesive Forces: These are molecular and electrostatic interactions at the skin-object interface that "stick" the two surfaces together.
- Deformation Forces: These forces result from the energy required to deform the compliant, layered structure of the skin as it conforms to the contacting surface.

Because human skin is nonlinear, viscoelastic, and anisotropic, its frictional behavior depends on the skin's mechanical properties and states. This complexity makes the overall friction force a combination of both adhesion and the deformation response of the skin.

In many conditions, especially on smooth, impermeable surfaces, the adhesive interactions dominate. For instance, when the skin is hydrated (e.g., when sweat occludes the micro-topographical features of the fingerprint ridges), the effective contact surface becomes smoother, leading to stronger adhesive bonds. This enhanced adhesion increases the friction force even at relatively low normal forces. Studies have highlighted that adhesion is a major contributor, with the deformation component sometimes assumed negligible in engineering contexts but not necessarily in the case of compliant skin [64, 77].

However, recent studies have shown that the micro-scale deformation of the skin can also contribute significantly to the total friction force [64]. As the skin deforms under a normal load, it adapts its contact profile by flattening and stretching. During sliding motions, the skin may exhibit complex deformation patterns (e.g., a "bow-wave" compression in front of a sliding probe and stretching behind it) that reflect its mechanical response. These deformations require energy, which adds to the friction force.

From a perceptual standpoint, the immediate cues of partial slip or stick arise from sudden changes in skin deformation, whether due to a loss of adhesive contact or an increase in deformational force. Indeed, the tactile system is highly attuned not to the absolute magnitude of the friction force, but to changes in the skin's deformation under load. For example, during the transition from a static state (sticking) to partial slip, the skin undergoes rapid, localized deformations such as sudden stretching or a change in contact area. These dynamic cues are detected by mechanoreceptors in the skin, which convert mechanical changes into neural signals. The rapid onset of deformation changes, resulting from a reduction in friction force due to a decrease in adhesion or a change in deformation behavior, alerts the nervous system for potential slip, prompting corrective grip adjustments.

Since the total friction force is a combination of adhesive and deformation contributions, the resulting skin deformation profile carries a complex signature. This signature includes the steady-state conditions (reflecting baseline adhesion) and the transient changes (reflecting dynamic deformation). The brain integrates these spatial and temporal patterns to assess the frictional quality of the contact. For instance, a smooth, well-adhered contact will generate a different deformation pattern compared to one where the skin is beginning to yield or stretch due to decreasing friction.

Changes in friction force influence how much grip force is required to maintain a secure hold. When the adhesive component decreases, e.g., if the skin dries or if the contact surface becomes smoother, the skin may exhibit greater deformation, signaling a potential risk of slip. This sensory feedback can lead to a reflexive increase in grip force, optimizing the balance between preventing slip and avoiding excessive gripping effort. Conversely, under high adhesion conditions, where friction force is high even at low normal forces, the skin deformation cues may indicate a secure contact, allowing for a lighter, gentler grip. Thus, friction perception is not solely determined by the measurable friction force but by the associated skin deformations that provide rich, dynamic information about the tactile interface.

Skin Stretch (Lateral Displacement) Skin stretch is among the most direct skin deformation cues for friction perception and slip detection. It refers to the lateral displacement or elongation of the skin's surface relative to its underlying structures when a tangential force is applied. In other words, as a force exerted in parallel to the skin, the surface of the finger pad is deformed, stretched out in the direction of the applied force. This phenomenon alters the distribution of strain across the skin's surface, effectively changing both the spatial pattern and the tension within the skin.

Skin stretch is a dynamic deformation that transforms the mechanical interaction at the finger-object interface into a rich sensory signal. When the skin is laterally displaced, the resulting deformation produces distinct spatiotemporal patterns of strain. Mechanoreceptors (especially slowly adapting afferents) are highly sensitive to these changes. They encode both the magnitude and the rate at which the skin is being stretched, thus providing real-time information about the frictional interaction between the skin and the contacting surface. Research shows that small amounts of imposed skin stretch (in the range of 0.25–0.75 millimeters) significantly increase the perceived friction, indicating even subtle lateral displacements can be detected and interpreted by the nervous systems [88]. An increase in stretch often correlates with more resistance at the interface (i.e., high friction), whereas a reduction can signal a slip or lower frictional grip. This distinction is key in detecting the transition between sticking and slipping. The dynamic changes in skin stretch also contribute to perceived stiffness. When the skin is stretched more, it signals that the skin-object interface is experiencing higher friction, which in turn can lead to an increased perception of stiffness.

The direction in which the skin is stretched relative to the natural kinesthetic force determines the mechanoreceptor firing and has notable effects on both perception and motor control. Mechanoreceptors in the fingertip, especially those slowly adapting afferents, exhibit preferred directional sensitivities [98]. This directional coding is crucial not only for detecting the onset of slip, but also for informing the brain about the vector along which the frictional forces are acting. Studies that compare the effects of positive versus negative skin stretch show that when artificial skin stretch is applied in the same direction as the natural kinesthetic force, it tends to augment the perceived stiffness and friction more robustly [28]. This alignment leads to higher grip forces as the tactile system interprets the deformation as an indicator of strong frictional engagement. Conversely, when the stretch is applied in the opposite direction, the augmentation of perceived stiffness is consistently lower. In this case, the grip force adjustments are more moderate, indicating that the nervous system interprets this directional mismatch as a less severe change in frictional conditions [28].

Contact Area Microscopic and macroscopic aspects of contact area also have significant implications for perceived friction. When the fingertip presses against a surface, only a portion of the apparent (visible) fingertip region actually makes contact at a microscopic level. This real contact area is formed by numerous micro-scale junctions where the skin's ridges and surface asperities interlock [52]. Because skin is nonlinearly compliant, this real contact area can expand or contract depending strongly on factors such as skin hydration, the applied normal load, and surface topography of both the skin and the object. Advanced methods including high-resolution imaging and ultrasonic techniques have been employed to precisely measure these microscopic contacts, which are directly linked to frictional strength.

Thus, the real contact area is not static. It evolves as the skin deforms under varying loads and tangential forces. This dynamic behavior transforms the contact area into a rich deformation cue that serves as a indicator of friction. Particularly, the friction force experienced at the skin-object interface is strongly governed by the real contact area. According to the adhesion friction model [17], a larger real contact area typically corresponds to stronger adhesive forces, which in turn increase the frictional resistance. As the skin deforms (e.g., when pressed or slide against a surface), the extent and distribution of these micro-contacts change, providing a direct measure of how much the skin is "gripping" the surface.

Conversely, if tangential forces rise enough to induce partial slip, such as during active manipulation or sliding, the contact area can decrease as tangential force increases particularly near the trailing edges, a phenomenon attributed to the nonlinear stiffening of the skin [14]. These dynamic alternations—from full contact to partial slip—result in characteristic skin deformation patterns. The central nervous system can use the evolution of these patterns, as measured by changes in contact area, to infer variations in friction. In essence, a reduction in the real contact area signals that the adhesive bonds are being broken or that local slip is beginning, thereby serving as an early-warning cue.

Moreover, distribution of force across the contact area is essential for realistic friction perception. Studies have demonstrated that simulating the actual deformation profile (i.e., matching the contact area and shape of the finger pad) yields more accurate friction perception than merely reproducing a single, lump-sum force [46]. This confirms that how the contact area is changed translates into critical tactile cues that the nervous system can interpret and form the perception of friction.

The influence of contact area on friction perception manifests through several interrelated effects:

- Determining Frictional Strength: Because the real contact area is directly tied to the number and strength of microscopic adhesive junctions, it plays a critical role in determining the CoF. Studies have shown that under different normal loads, variations in the contact area correlate with changes in the measured CoF [18]. For example, under lower contact pressures, an increased real contact area (enhancing adhesion) results in higher friction, whereas a decrease in contact area with increasing pressure can reduce the CoF.
- Providing Spatial-Temporal Information: The distribution of the contact area over the finger pad rise to a spatial map of stress and strain. As the finger interacts with a surface, changes in the contact area can be sensed by mechanoreceptors distributed across the skin. These receptors detect not only the magnitude of local deformation, but also their spatial patterns. A rapidly shrinking or shifting in contact area can indicate the onset of slip. Such temporal changes in contact area serve as a cue for both the perception of friction and for motor responses.
- Slip Detection and Grip Control: Changes in contact area, especially localized shrinkage or shift, alert the somatosensory system for potential slip. Mechanoreceptors tuned to skin deformation detect if the contact patch is losing coverage, prompting reflexive or voluntary increases in grip force to restore frictional safety margins. The high correlation between real contact area and frictional strength makes contact-area-related feedback crucial for tasks requiring stable manipulation [52].

In summary, contact area is more than a geometric artifact; it is a skin deformation cue that underlies how friction is both physically generated and perceived. By dynamically reflecting the interplay between adhesive forces and skin deformation, variations in contact area provide the nervous system with a comprehensive signal about the state of the interface, informing about grip security and slip risk.

Contact Shape Contact shape refers to the geometry of the skin-object contacting interface. The geometry or shape of the contact region determines the distribution of internal mechanical stresses in the skin. As the skin deforms under load and during the onset of shear forces, the shape of contact changes (e.g., from circular to more irregular or elliptical patterns). Variations in surface curvature, for example, a flat vs. a curved object, lead to different deformation profiles in the fingertip [54]. Changes in contact shape alter the spatial gradients of strain across the fingertip. These gradients are critical for the spatial coding of tactile stimuli, as different regions of the contact surface may experience varying degrees of deformation. Non-uniform contact shapes can also localize high-stress regions that are prone to micro-slip, eliciting abrupt shifts in skin strain and thus bursts of neural activity. Research illustrates that the evolution of the contact shape is closely linked to the dynamics of skin deformation, providing rich spatial information that is encoded by tactile afferents during the transition from full adhesion (stick phase) towards partial slip [14].

Vibration Vibration is a powerful mechanical cue that contributes to friction perception by conveying dynamic information about the interaction between the skin and a surface. When a fingertip slides across a surface—even one that is seemingly smooth—the microscopic asperities on the skin and on the surface create small oscillations. These oscillations can vary in amplitude and frequency based on the factors such as surface microgeometry (e.g., presence of microscopic protrusions), contact speed and force (higher sliding speed often leads to higher-frequency vibrations), and frictional and adhesive properties (e.g., surface moisture, compliance). Essentially, as the skin deforms and these contact-induced vibrations propagate through the skin, these dynamic events create a spectrum of vibratory signals that reflect the frictional properties of the surface.

The vibrations generated at the skin-object interface cause rapid, transient deformations in the skin tissue. Rapidly adapting afferents are highly sensitive to these high-frequency signals. As these receptors transduce the vibratory energy into electrical impulses, they provide the central nervous system with detailed information about the frictional conditions of the surface. Vibration plays a key role in detecting slip events. Small vibration can precede or accompany micro-slip at the periphery of the contact [4]. Even before a full slip occurs, localized slip events, often undetectable to the naked eye, produce high-frequency vibrations. Thus, these signals serve as an early warning system, prompting the central nervous system for motor adjustments.

Vibration can be used to encode surface properties and microgeometry. According to the study using frequency analysis technique, the frequency content of the vibration spectrum is influenced by characteristic contact parameters, such as scanning speed, surface roughness, and texture [26]. This means that subtle differences in the way a surface interacts with the skin (e.g., whether the surface is smooth or textured) are reflected in distinct vibratory patterns. The nervous system decodes these patterns using them to differentiate between materials and to judge frictional sensations.

Importantly, vibration-based mechanisms work alongside deformation-based cues [4]. While deformation cues (e.g., changes in contact area or skin stretch) provide information about skin state, the vibratory signals offer high-frequency, fine-grained data that enhance the resolution of friction perception. Together, they form a complementary and robust system for sensing changes in friction, shaping both our subjective experience (e.g., slipperiness) and physical interactions (e.g., grip control).

Friction perception is the neural interpretation of dynamic changes in skin deformation, serving as a critical mechanism for grip control and the perception of stickiness, slipperiness, and dryness. Unlike a simple force measurement, it is the transition in the skin's deformation state—signaling the shift from stick to slip—that the nervous system detects most acutely. This percept is built from the integration of several key mechanical and vibratory cues generated at the skin-object interface. Defining aspects of friction perception include:

- **Skin Stretch:** Lateral displacement of the skin is a primary cue. Even sub-millimeter stretch provides direct information about frictional resistance and its direction, with aligned stretch increasing perceived stickiness and stiffness.
- **Contact Area Dynamics:** The real contact area, governed by adhesion, is a powerful indicator. An expanding area signals secure grip and high friction, while a rapidly shrinking or shifting area is a critical early-warning signal for slip.
- **Coefficient of Friction (CoF):** For human skin, the CoF is not constant. It is adhesion-dominated at low forces (causing a "sticky" feel) and decreases at higher loads, allowing the nervous system to fine-tune grip force efficiently.
- **Vibration:** High-frequency vibrations generated by micro-slip events and surface asperities are encoded by RA and PC afferents, providing an early warning system for slip long before a full macro-slip occurs.
- **Contact Shape:** The geometry of the contact region determines internal stress distribution. Evolving contact shapes during shear provide rich spatial information about the transition from full adhesion to partial slip.

The brain integrates these dynamic signals—the spatial map of contact, the rate of skin stretch, and the spectral content of vibrations—to form a real-time assessment of frictional stability. This allows for the precise motor adjustments that characterize dexterous manipulation, ensuring we can grip objects securely without excessive force.

2.6.4 Shape: Recognition of Edges, Curves, and Contours

Shape perception unifies the mechanical and geometric dimensions of touch; it depends not just on deformation, but also on how multiple contacts integrate over time. Human tactile perception is a multifaceted process that relies on several key dimensions, including roughness, compliance, friction, and shape. Each of these dimensions corresponds to different physical properties of objects, and our sensory system has evolved specialized mechanisms to detect these properties. Extending beyond sensing material-based properties like compliance, the perception of shape enables us to understand an object's spatial structure and geometric features through the detection of edges and curvature. These features serve as fundamental building blocks for constructing a mental representation of objects' physical form through touch: edges define the boundaries that outline an object's form, while the orientation of these edges supplies directional information critical for detecting contours and aligning the grip appropriately. Curvature, on the other hand, conveys nuanced details that differentiate between flat, rounded, or concave surfaces, thereby offering insight into the intrinsic shape characteristics of the object. The global perception of an object's shape emerges from the integration of these discrete geometric features, and this ability forms the foundation for critical tasks like object manipulation, grasping, and object recognition. Without the capacity to detect an object's boundaries, contours, and form, the nervous system would struggle to generate a coherent representation of objects, making fine motor control and dexterous interactions significantly more difficult.

Shape perception presents an added layer of complexity compared to the other tactile dimensions. While the initial detection of local features such as edges and curvature is critical, the integration of these geometric cues is essential for constructing a global representation of an object's form. In this context, the nervous system does not treat edges or curvature as isolated phenomena. Instead, it synthesizes this information to generate a comprehensive shape percept. Unlike other tactile perceptions, such as compliance or friction, that can be sensed instantaneously at a single contact point via localized activity of mechanoreceptors in the skin, shape perception requires linking multiple touchpoints across the surface and across time, leading to more elaborate, distributed neural processing across several brain regions. The initial sensory input captured by the mechanoreceptors is transmitted to the brain, where higher-order cortical areas come into play. These cortical regions integrate the various tactile signals, both cutaneous (from the skin) and proprioceptive (information about limb position and movement), to build a coherent representation of an object's shape. This integration process involves complex computations that cannot be accomplished by local neural responses alone, highlighting the role of distributed cortical networks in tactile shape perception.

Detection of Edge and Edge Orientation Edge detection refers to the process by which tactile receptors in the skin identify discontinuities in a surface, such as sharp boundaries, ridges, or edges of objects. Edge orientation detection, on the other hand, involves determining the directional alignment of these edges relative to the skin or the hand's movement. Both processes are fundamental to haptic perception, allowing humans to recognize and manipulate objects effectively. Studying edge and their orientation provides insight into how the nervous system encodes spatial and geometric features through mechanoreceptor activity and cortical processing, which can be translated into practical applications for improved sensing and feedback capabilities.

Stress Concentration and Mechanoreceptor Activation The skin's microstructure—its papillary ridges and the multiple, spatially segregated mechanoreceptive sites they contain—plays a critical role in how stress and strain are distributed when an edge contacts the fingertip. When an edge contacts the skin, it deforms the compliant tissue, generating localized regions of high stress (and corresponding strain) particularly at the ridge tips. The intermediate ridges amplify and guide these mechanical forces towards clusters of mechanoreceptors [30]. This modulation enhances the spatial precision of edge detection by increasing the sensitivity of SA-I afferents (i.e., Merkel Cells), which provide high-resolution spatial detail, and enhancing the responsiveness of RA-I afferents (i.e., Meissner corpuscles) during dynamic contact (e.g., onset or offset), ensuring quick feedback when an edge moves across the skin [29, 55].

Edge Sharpness and Deformation Response The shape of an edge (sharp vs. blunt) affects how it deforms the skin. A sharper edge creates a more distinct pressure gradient, leading to stronger activation of local afferent populations. The interplay between stress concentration and mechanoreceptor density enables the fingertip to extract fine spatial details about the edge's properties. Essentially, the skin converts these geometric differences into measurable mechanical-electrical signals that the peripheral nervous system can encode.

Both firing rate intensity and temporal sequencing contribute to edge orientation perception, operating in independent and complementary ways. The nervous system exploits both mechanisms simultaneously, where the intensity signal provides a coarse estimate of edge orientation at static and dynamic contact, while the temporal signal refines this estimate with enhanced precision, especially during dynamic touch. These mechanisms ensure that by the time signals reach the somatosensory cortex, much of the feature extraction work is already done. Cortical neurons in area 3b and 1 of S1 and neurons in S2 then refine and integrate these signals for higher-order shape perception [42, 89, 100].

Encoding of Surface Curvature Surface curvature is a fundamental aspect of tactile shape perception, and it is achieved through both the precise patterning of responses from peripheral mechanoreceptors and subsequent central processing. SA-I afferents, in particular, play a crucial role in encoding surface curvature; their responses are shaped by both the intensity of activation and the spatial distribution of active afferents, which vary depending on the curvature of a stimulus.

Peripheral Encoding: Population Responses from Mechanoreceptors When a curved surface contacts the fingertip, the deformation of the skin produces a spatial pattern of stress that is transduced by these afferents. Specifically, scientific findings demonstrated that SA-I afferents increase their firing rates as curvature increases [32, 60]. This occurs because highly curved surface creates steeper gradients of skin deformation, which lead to greater local stress concentrations in the receptive fields of these afferents. The firing rates are directly proportional to both the magnitude of curvature and the contact force [33]. For example, when a spherical stimulus with a smaller radius (i.e., higher curvature) is applied, the response profile of a given SA-I becomes both higher in amplitude and narrower in width compared to responses evoked by a flatter surface. This spatial profile—essentially the "shape" of the response across the receptive field—mirrors the curvature of the object. The study that applied spherical surfaces with a range of curvatures showed that the response profiles of SA-I afferents consistently scaled with curvature, while the responses of SA-II afferents saturate with higher curvature—but with contact force further modulating the response magnitude [33].

While both afferent types respond to sustained pressure, the distinction between the functional roles of SA-I and SA-II mechanoreceptors lies in their receptive field properties and response characteristics to changes in stimulus position, curvature, and contact force. SA-I afferents are densely packed in the fingertips and possess small, well-defined receptive fields with multiple sensitive zones, which makes them highly suited for detecting fine spatial details of an object's surface. When a stimulus is moved across the receptive field of an SA-I, the response shifts accordingly (e.g., peaks, widths), maintaining a high-resolution representation of the contact location. SA-II afferents, in contrast, have larger receptive fields with less spatially defined sensitive zones, making them better suited for global force perception but less effective for detailed spatial discrimination. Unlike SA-I afferents, which linearly increase their response with increasing curvature, further increases in curvature do not elicit stronger responses in SA-II afferents once a certain curvature threshold is reached. However, SA-II afferents exhibit a proportionally larger response to increase contact force than SA-I afferents, making them more effective for detecting force fluctuations. In a distinct but complementary way, SA-I afferents provide fine spatial resolution for surface curvature and contact position encoding, whereas SA-II afferents excel in detecting force changes but offer a coarse representation of local shape of contacting surface.

In contrast to SA afferents, RA afferents (both RA-I and RA-II) show much smaller responses when a static curved stimulus is applied. Their firing does not systematically vary with changes in curvature or contact force, implying that they are less involved in encoding the fine details of surface curvature [32, 33]. However, the RA afferents may be more involved than SA afferents in signaling dynamic changes, such as the onset of contact or the movement of a curved edge during stroking, where transient information is crucial [66]. Nonetheless, for the detailed encoding of curvature, the SA afferents are the primary carriers of information.

2.6.4.1 Effects of Cutaneous Cues on Curvature Discrimination

Cutaneous cues play a critical role in the discrimination of surface curvature by providing detailed information about the way a stimulus deforms the skin. Passive touch experiments—where curved stimuli were pressed against immobilized finger pads—were conducted to isolate the contribution of cutaneous cues from any kinesthetic influences [59]. This approach demonstrated that when only cutaneous inputs are available, humans can discriminate differences in curvature as small as a 10% change (a Weber fraction of 0.1). The skin's deformation profile, in particular the spatial gradient or local slope across the contact area, provides essential cues that the nervous system uses to infer the curvature of the surface.

Moreover, studies have revealed a strong correlation between the contact area and the perceived curvature [31]. As the curvature of a stimulus increases, the contact area between the object and the skin changes in a predictable manner, leading to a distinct spatial distribution of pressure and strain across the fingertip. This change in contact area results in different activation patterns in the slowly adapting (SA) afferents with higher in intensity and more focused, whose response profile has both intensity-based and spatial-based mechanisms. Such response profiles provide a neural "signature" for high curvature that can be discriminated reliably. Importantly, when the contact area was held constant, the discrimination thresholds remained similar, indicating that it is not the absolute extent of skin stimulation, but rather the gradient of deformation–the differences in local attitude across the contact region–that is critical for curvature perception. This is further supported by studies using active touch conditions, which find that increasing the contact area leads to better discrimination performance because it generates a larger overall deformation gradient [34].

These behavioral findings are further supported by neural recordings. In studies involving both monkey and human subjects, SA afferents were found to encode curvature robustly, with their firing rates increasing in response to greater curvature and stronger contact forces [32, 33]. This response scaling means that the population activity of these receptors reflects not only the magnitude of the deformation but also its spatial distribution, effectively encoding the geometric features of the stimulus. On the other hand, rapidly adapting (RA) afferents contributed little to curvature discrimination in these static, passive touch conditions, underscoring the primary role of cutaneous SA afferents in this perceptual task.

Another important factor in curvature discrimination is contact length, or the extent of the curved surface in contact with the skin. Studies found that longer contact lengths decrease curvature discrimination thresholds, as they allow for a greater portion of the curvature gradient to be processed by mechanoreceptors [86]. This finding aligns with the previous work where larger contact areas led to better performance, reinforcing the idea that curvature discrimination relies on detecting local slope changes across the stimulus contact area [87]. This is consistent with the hypothesis of curvature perception, which states that the first-order information (i.e., differences in local surface attitude) provide the primary cues for tactile curvature discrimination, whereas zeroth-order information (i.e., height or aspect ratio) is not sufficient for determining discrimination thresholds. The fact that removing second-order cues (i.e., curvature) from the stimuli does not significantly affect discrimination thresholds further validates the idea that the tactile system relies on gradual changes in skin deformation rather than precise encoding of absolute curvature values [110].

Furthermore, curvature discrimination is largely independent of bimanual vs. unimanual exploration. Studies that examined perceptual performance across different fingers and hands found that the discrimination thresholds remained similar regardless of whether stimuli were presented to one or both hands [92]. This suggest that curvature perception is localized to the contact region. However, individual biases in curvature perception were observed, likely due to small asymmetries in hand positioning or slight differences in mechanoreceptor sensitivity between fingers.

Overall, the integration of these cutaneous cues—changes in contact area, local deformation gradients, and resulting activities of SA afferent responses—contributes to our ability to accurately discriminate surface curvature. These local signs are interpreted centrally, where they form a holistic representation of object shape. Such precise encoding of curvature is essential not only for recognizing the form of objects, but also for guiding the fine motor actions, like gripping or manipulating objects, where understanding subtle variations in surface geometry can be crucial for performance.

In short, the perception of curvature is a fundamental building block for recognizing complex object shapes. Objects in the environment often contain multiple curved surfaces, and our ability to detect curvature allows us to infer contour continuity, object edges, and fine surface details. Curvature perception allows the brain to decompose complex shapes into simpler, more manageable components. Similar to how edge orientation helps define object contours in vision, curvature detection defines surface geometries in touch. For example, distinguishing between a sphere, a cylinder, and a flat plate relies on detecting how surface curvature changes across contact areas. Ultimately, the perception of curvature contributes to the broader perception of 3D shape, object recognition, and dexterous manipulation, making it an essential aspect of tactile sensory function.

2.6.4.2 Perception of Object's Shape

The detection of edges, their orientation, and local surface curvature forms the foundation for our tactile perception of shape. Edges provide the basic boundaries of an object, while their orientation specifies the direction of these boundaries. Curvature adds further details by indicating how these edges bend—distinguishing, e.g., convexity from concavity. In combination, these basic features enable the nervous system to construct a detailed representation of an object's shape from the myriad of local tactile cues.

Roles of SA and RA afferents in Shape Perception The encoding of shape through cutaneous cues is achieved by an intricate interplay between SA and RA afferents, each contributing distinct but complementary information to the overall percept of an object's form. SA afferents, particularly those innervating the finger pad, provide sustained responses that closely track the geometric features of a stimulus. When a stimulus with a stepped or wavy profile is indent into the skin, SA afferents generate a spatial response profile that reflects local variations in curvature [67]. Studies showed that SA afferents exhibit the highest firing rates over convex (steep) regions of a step, reduced firing over concave regions, and intermediate rates over flat regions [97]. This graded, steady response, observed both during the dynamic ramp phase and the static phase, enable a precise mapping of local curvature. Moreover, when stimuli (e.g., wavy surfaces) are stroked across the finger pad, the averaged spatial rate profiles of SA afferent responses reveal invariant features, such as constant slopes along the rising and falling edges of a burst, which can serve as a neural code for the constant curvature of an object. These robust and stable SA signals are critical for encoding fine details of shape including the orientation of an edge, as the spatial pattern of activity across a population of SA afferents closely mirrors the two-dimensional (2D) "footprint" of the stimulus.

In contrast, RA afferents respond primarily during the dynamic phase of stimulation. Their responses are characterized by brief bursts of activity that are highly sensitive to the onset and rate of skin deformation rather than to static curvature details. RA responses are strongly influenced by velocity, meaning that as the speed of stroking increases (e.g., 1 to 40 mm/s), the number of activated RA afferents increases and their burst rates become higher. However, unlike SA responses, RA bursts tend to be less spatially modulated; they provide a more transient, less detailed signal about local surface curvature. Although RA signals do contribute to the early detection of contact (especially contact location) and rapid changes in skin deformation, their limited spatial modulation means that they play a subordinate role in the precise discrimination of curvature and shape.

The Role of Skin Mechanics in Shape Perception The mechanical properties of the skin play an important role in shaping how tactile stimuli are encoded by mechanoreceptors and ultimately perceived as object shape. When an object makes contact with the skin, the applied forces do not create a direct imprint; instead, they propagate through the compliant layers of the skin, generating a complex pattern of stresses and strains. Such mechanical deformation depend both on the magnitude and distribution of the contact force as well as on the inherent physical properties of the skin. Studies describe that the deformation of the skin creates a somewhat distorted neural image of the actual shape of the object [85, 99]. This distortion arises due to factors such as the elasticity of the skin, the thickness of the epidermis, and the distribution of underlying mechanoreceptors. Models that convert the displacement profile of an indented spatial pattern into a stress and strain profile show that the skin tends to enhance certain features (e.g., edges and corners) while attenuating others (e.g., smooth continuous surfaces).

The angle and direction of contact are also critical factors because they influence the way the skin deforms and the signals are transmitted. Depending on whether the force is applied perpendicularly or at an angle, the resulting strain distribution across the skin can vary dramatically. For example, a force applied obliquely might cause more shear deformation compared to a purely normal force, affecting not only the magnitude of the stress reaching the mechanoreceptors but also altering the spatial filtering properties of the skin [72]. The direction of force application also significantly affects how shape information is perceived by mechanoreceptors. The responses of mechanoreceptive afferents are not isotropic, meaning they do not respond equally to forces applied from different directions. Studies found that tactile afferents in the fingertips can reliably encode both shape and force direction [7, 55]. These force direction cues are essential for interpreting object curvature and orientation, allowing the brain to determine both the shape as well as also how the object is oriented relative to the skin.

- The Interplay of Skin Properties and Mechanoreceptors: The physical properties of the skin, including its mass, stiffness, and resistance, interact with the characteristics of mechanoreceptors to determine shape perception. Different types of mechanoreceptors are embedded in the skin at different depths and have unique receptive field properties that influence how they encode shape-related information. Particularly, SA-I afferents, having small, well-defined receptive fields with multiple highly sensitive "hot spots" respond to static indentations and convey high-resolution spatial information, making them particularly adept at signaling precise features like edge orientation and local curvature. SA-II afferents respond to skin stretch and encode information about global shape and force direction. They have large receptive fields, which makes them poor at encoding fine details but excellent at detecting overall shape deformation and curvature changes. In contrast, RA-I afferents respond transiently and generally have larger receptive fields. They are not tuned for static shape perception, but play an essential role in detecting shape through movement. For example, when a hand strokes a surface, RA-I afferents capture transient changes in surface curvature and edge discontinuities.
- Selectivity for Spatial Features: Mechanoreceptors are selectively tuned to different spatial features, including edges, curvature, and orientation. This selectivity arises from the combined effect of skin mechanics and the unique way in which mechanoreceptors distribute their sensitivity zones across the skin. Edge detection and orientation sensitivity are achieved due to the SA-I afferents' high sensitivity to edges and discontinuities, with their receptive fields responding most strongly at points of high curvature or sudden changes in surface orientation [6, 89]. Many neurons in areas 3b and 1 of the S1 cortex respond selectively to specific edge orientations, similar to orientation-selective neurons in the visual cortex [91].

  Neurons in area 2 of the S2 cortex are specifically tuned to detect curvature of contours [116, 117]. As an object's curvature increases, SA-I afferents produce higher firing rates in response to convex surfaces and lower firing rates to concave surfaces, emphasizing high-contrast features. These signals serve as the foundational input for further processing in the central nervous system, where higher-order areas refine these inputs into invariant percepts of shape.

In summary, shape perception emerges from a hierarchical process that begins with the transformation of applied forces by the skin's mechanics. The direction of contact force influences how mechanoreceptors encode curvature, edges, and orientation, while the physical properties of the skin determine how forces are distributed and how effectively key features are highlighted. Different mechanoreceptor types contribute unique spatial features to shape perception and the feature-selectivity of these receptors supports a progression of information refined from peripheral regions to cortical processing areas to construct a coherent representation of object shape.

This process is built upon the integration of several fundamental cues and mechanisms:

- **Edge and Orientation Detection:** The skin's microstructure, particularly the papillary ridges, amplifies stress at contact points, creating distinct neural activity patterns. SA-I afferents provide high-resolution spatial maps of edges, while the temporal sequence of activation across receptors during movement provides critical information about an edge's orientation.
- **Curvature Encoding:** SA-I afferents are the primary encoders of surface curvature, with their firing rates and spatial response profiles scaling directly with the degree of curvature. The local deformation gradient across the contact area is a more critical cue than the absolute contact area itself.
- **Mechanoreceptor Specialization:** The nervous system employs a division of labor where SA afferents provide the sustained, high-fidelity spatial details for static shape, and RA afferents contribute dynamic information about movement and contact transients, which is essential for shape perception through active exploration.
- **Cortical Feature-Selectivity:** Peripheral signals are refined in the somatosensory cortex, where neurons become selectively tuned to specific features, such as edge orientation in areas 3b and 1, and contour curvature in areas 2 and S2, mirroring the functional hierarchy seen in visual processing.
- **Spatial Integration:** Unlike material properties, global shape perception requires the integration of multiple local geometric features (edges, curvature) across space and time, forming a coherent representation of an object's 3D form from discrete touchpoints.

Put simply, the perception of an object's shape is a constructive process, where the brain transforms low-level mechanical deformations of the skin into increasingly complex and invariant geometric representations, enabling dexterous manipulation and object recognition.

2.7 Touch as a Multi-Modal and Integrative Sense

The cue–dimension mappings from Sect. 2.6 provide a first-principles bridge from mechanics to psychophysics, however, the cues are not processed in isolation. To achieve complex tasks that demand dynamic adjustment and higher-order cognition, such as dexterous manipulation or social communication, the human sense of touch operates as a multi-modal and integrative process. Rather than relying on a single sensory channel, touch emerges from the coordination of multiple subsystems that together enable humans to perceive, interpret, and interact with both the physical and social world. The somatosensory system sits uniquely at the interface between body and environment, constantly converting mechanical interactions into perceptual and emotional meaning.

For example, touch integrates cutaneous and kinesthetic feedback to support discriminative and manipulative tasks. The skin encodes fine-scale contact cues such as pressure, vibration, and texture, while proprioceptive receptors in muscles and joints provide continuous information about movement, position, and applied force. Working together, these modalities allow for adaptive control, stable grasping, and precise coordination—capabilities that no single feedback source can achieve alone.

Beyond its functional role, touch also mediates affective and social communication. Gentle, slow stroking of the skin project to emotion-related brain regions and evoke feelings of comfort, empathy, and trust. Through these pathways, touch becomes a means not only of sensing the external world but also of expressing and perceiving human connection.

2.7.1 Integration of Cutaneous and Kinesthetic Feedback

Although kinesthetic and cutaneous feedback arise from distinct sensory pathways, they are fundamentally interdependent and complementary. Their integration provides the nervous system with both the global context of body posture and force (kinesthetic) and the localized detail of skin deformation (cutaneous) [9, 111]. This integration enables seamless and adaptive responses to environmental changes [73]. For tasks demanding precision and dexterity, such as threading a needle, assembling intricate components, or regulating grip force during object manipulation, both channels must be functional [84]. When either system is impaired, through injury, neurological conditions, or sensory loss, performance degrades significantly, underscoring the necessity of their cooperation.

At the neural level, integration occurs across multiple stages of the somatosensory hierarchy [10]. Peripheral mechanoreceptors encode surface features, while proprioceptors track joint and muscle states. In the primary somatosensory cortex (S1), convergent inputs begin to form representations of orientation and curvature [51]. Higher-order cortical areas (e.g., S2 and area 2 of S1) combine cutaneous and proprioceptive signals, enabling stable 3D shape perception even as the hand moves or changes posture [61, 118]. This integration follows principles of reweighted cue combination: when surface curvature cues are more robust, the brain relies more heavily on cutaneous input; when postural information is more reliable, proprioception dominates. Such dynamic weighting ensures optimal perception across diverse contexts.

For example, the necessity of cutaneous-proprioceptive integration for shape discrimination has been experimentally validated. Studies that investigated how human perceive differences in 2D angles using either cutaneous feedback from the fingers or proprioceptive feedback from the shoulder found that, when only proprioceptive input was available (via active shoulder movement), discrimination thresholds increased significantly, indicating that cutaneous input is critical for fine shape perception [106]. Likewise, when only cutaneous feedback was available (via passive touch), thresholds also increased, suggesting that proprioceptive cues also contribute to discrimination shape details. When both sensory modalities were removed, participants could not discriminate shape at all, highlighting the necessity of multisensory integration for accurate shape perception. Similarly, in examining how force and positional cues contribute to shape perception, studies found that when participants actively explored bumpy surfaces, both skin deformation and the finger's movement trajectory influenced perceived curvature [21]. Interestingly, the weight of these cues depended on surface geometry: for more convex surfaces, position cues dominated, while for flatter surfaces, force cues carried more weight, which suggests that the brain assigns greater weight to the more reliable cue in a given context, ensuring optimal shape perception.

2.7.2 *Affective Touch and Social Communication*

The impact of touch is deeply embedded in human behavior. Beyond facilitating manipulation and object exploration, touch also plays a critical role in social interaction and emotional communication, serving as a powerful and direct medium for conveying emotions and social cues.

Studies have demonstrated that even brief touches can accurately convey emotions such as fear, love, or angle—often without the need for visual or auditory cues [44, 45]. The underlying mechanism is the C-tactile (CT) afferents—a class of low-threshold mechanoreceptors that are especially responsive to slow, gentle touches [35, 94, 107]. When activated, these receptors engage neural pathways linked to emotional processing and social bonding.

Cutaneous feedback further reinforces emotional and alert signal transformation. Subtle mechanical cues—such as the degree of skin indentation and stretch, the speed of contact, and the force profile—play a significant role in the perception of emotional states [114]. For instance, a light, lingering touch may convey calmness and affection, whereas a sudden, firm grasp could signal urgency or fear [115]. By translating mechanical touch behavior into nuanced emotional content, these tactile signals form a complex, non-verbal communication channel that is essential for establishing trust, encouraging cooperation, and deepening social bonds [41].

Moreover, kinesthetic feedback contributes to social touch by guiding the force and motion of physical interactions. During social gestures like a handshake or a hug, kinesthetic control ensures appropriate force application, while cutaneous feedback refines the sensation of contact, enabling the interpretation of subtle social cues. This delicate interplay highlights the integral role of touch in affective communication and interpersonal connection, although often silent and subconscious; it profoundly shapes how we connect with one another and maintain lasting relationships [78, 101].

This multi-modal and integrative character of touch underscores why haptics must be studied and engineered as a system rather than as a collection of isolated components. Human touch does not arise from any single sensory channel or neural mechanism; it emerges from the coordinated interaction of biomechanics, neural processing, and perception. In the biological system, tactile perception depends on how mechanical deformation of the skin activates receptor populations, how those neural signals are integrated across spatial and temporal scales, and how the brain interprets them in context. Similarly, in an artificial system, true haptic fidelity cannot be achieved by optimizing individual parts—sensors, actuators, or control algorithms—in isolation. Instead, it requires understanding and reproducing the causal chain that links physical contact, mechanical deformation, neural encoding, and perceptual outcome.

2.8 Conclusion

Unveiling the linkages between skin mechanics and psychophysics provides the scientific foundation for a first-principles approach to haptic design. It allows system goals to be derived not empirically or heuristically, but from measurable, biologically grounded relationships. By quantifying how deformation cues correspond to perceptual experiences, one can define hardware design targets—such as sensitivity, spatial resolution, or force bandwidth—that mirror the limits of human tactile sensitivity, and software design goals—such as signal encoding, feature extraction, or feedback control—that emulate the organization of neural coding in a manner tailored to suit that end-use application in the broader system. Viewing haptics through this systems lens ensures that artificial touch is not just functionally effective, but perceptually meaningful: aligned with the same physical and neural logic that gives rise to the human sense of touch itself.

References

1. Abraira, V.E., Ginty, D.D.: The sensory neurons of touch. Neuron **79**(4), 618–639 (2013)
2. Ackerley, R., Watkins, R.H.: Microneurography as a tool to study the function of individual c-fiber afferents in humans: responses from nociceptors, thermoreceptors, and mechanoreceptors. J. Neurophysiol. **120**(6), 2834–2846 (2018)

3. Barrea, A., Bulens, D.C., Lefevre, P., Thonnard, J.L.: Simple and reliable method to estimate the fingertip static coefficient of friction in precision grip. IEEE Trans. Haptics **9**(4), 492–498 (2016)
4. Barrea, A., Delhaye, B.P., Lefèvre, P., Thonnard, J.L.: Perception of partial slips under tangential loading of the fingertip. Sci. Rep. **8**(1), 7032 (2018)
5. Bell-Krotoski, J., Weinstein, S., Weinstein, C.: Testing sensibility, including touch-pressure, two-point discrimination, point localization, and vibration. J. Hand Therapy **6**(2), 114–123 (1993)
6. Bensmaia, S.J., Denchev, P.V., Dammann, J.F., Craig, J.C., Hsiao, S.S.: The representation of stimulus orientation in the early stages of somatosensory processing. J. Neurosci. **28**(3), 776–786 (2008)
7. Birznieks, I., Jenmalm, P., Goodwin, A., Johansson, R., et al.: Directional encoding of fingertip force by human tactile afferents. Acta Physiologica Scand. **167**(2), A24 (1999)
8. Blake, D.T., Hsiao, S.S., Johnson, K.O.: Neural coding mechanisms in tactile pattern recognition: the relative contributions of slowly and rapidly adapting mechanoreceptors to perceived roughness. J. Neurosci. **17**(19), 7480–7489 (1997)
9. Blanchard, C., Roll, R., Roll, J.P., Kavounoudias, A.: Combined contribution of tactile and proprioceptive feedback to hand movement perception. Brain Res. **1382**, 219–229 (2011)
10. Bodegård, A., Geyer, S., Grefkes, C., Zilles, K., Roland, P.E.: Hierarchical processing of tactile shape in the human brain. Neuron **31**(2), 317–328 (2001)
11. Boundy-Singer, Z.M., Saal, H.P., Bensmaia, S.J.: Speed invariance of tactile texture perception. J. Neurophysiol. **118**(4), 2371–2377 (2017)
12. Connor, C.E., Johnson, K.O.: Neural coding of tactile texture: comparison of spatial and temporal mechanisms for roughness perception. J. Neurosci. **12**(9), 3414–3426 (1992)
13. Curray, J.R.: The analysis of two-dimensional orientation data. J. Geol. **64**(2), 117–131 (1956)
14. Delhaye, B., Lefevre, P., Thonnard, J.L.: Dynamics of fingertip contact during the onset of tangential slip. J. R. Soc. Interface **11**(100), 20140698 (2014)
15. Delhaye, B.P., Jarocka, E., Barrea, A., Thonnard, J.L., Edin, B., Lefèvre, P.: High-resolution imaging of skin deformation shows that afferents from human fingertips signal slip onset. eLife **10**, e64679 (2021)
16. Delhaye, B.P., Schiltz, F., Crevecoeur, F., Thonnard, J.L., Lefèvre, P.: Fast grip force adaptation to friction relies on localized fingerpad strains. Sci. Adv. **10**(3), eadh9344 (2024)
17. Derler, S., Gerhardt, L.C.: Tribology of skin: review and analysis of experimental results for the friction coefficient of human skin. Tribol. Lett. **45**, 1–27 (2012)
18. Derler, S., Gerhardt, L.C., Lenz, A., Bertaux, E., Hadad, M.: Friction of human skin against smooth and rough glass as a function of the contact pressure. Tribol. Int. **42**(11-12), 1565–1574 (2009)
19. Dhong, C., Miller, R., Root, N.B., Gupta, S., Kayser, L.V., Carpenter, C.W., Loh, K.J., Ramachandran, V.S., Lipomi, D.J.: Role of indentation depth and contact area on human perception of softness for haptic interfaces. Sci. Adv. **5**(8), eaaw8845 (2019)
20. Di Luca, M., Ernst, M.O.: Computational aspects of softness perception. In: Multisensory softness: perceived compliance from multiple sources of information, pp. 85–106. Springer, Berlin (2014)
21. Drewing, K., Ernst, M.O.: Integration of force and position cues for shape perception through active touch. Brain Res. **1078**(1), 92–100 (2006)
22. Dzidek, B.M., Adams, M.J., Andrews, J.W., Zhang, Z., Johnson, S.A.: Contact mechanics of the human finger pad under compressive loads. J. R. Soc. Interface **14**(127), 20160935 (2017)
23. Edwards, C., Marks, R.: Evaluation of biomechanical properties of human skin. Clin. Dermatol. **13**(4), 375–380 (1995)
24. Egner, T., Hirsch, J.: Cognitive control mechanisms resolve conflict through cortical amplification of task-relevant information. Nature Neurosci. **8**(12), 1784–1790 (2005)
25. Ellingsen, D.M., Leknes, S., Løseth, G., Wessberg, J., Olausson, H.: The neurobiology shaping affective touch: expectation, motivation, and meaning in the multisensory context. Front. Psychol. **6**, 1986 (2016)

26. Fagiani, R., Massi, F., Chatelet, E., Berthier, Y., Akay, A.: Tactile perception by friction induced vibrations. Tribology Int. **44**(10), 1100–1110 (2011)
27. Fakhoury, E., Culmer, P.R., Henson, B.: The effect of indentation force and displacement on visual perception of compliance. In: 2015 IEEE World Haptics Conference (WHC), pp. 88–93. IEEE (2015)
28. Farajian, M., Leib, R., Kossowsky, H., Nisky, I.: Direction-specific effects of artificial skin-stretch on stiffness perception and grip force control. IEEE Trans. Haptics **16**(4), 836–847 (2023)
29. Gerling, G.J.: Sa-i mechanoreceptor position in fingertip skin may impact sensitivity to edge stimuli. Appl. Bionics Biomech. **7**(1), 19–29 (2010)
30. Gerling, G.J., Thomas, G.W.: The effect of fingertip microstructures on tactile edge perception. In: First Joint Eurohaptics Conference and Symposium on Haptic Interfaces for Virtual Environment and Teleoperator Systems. World Haptics Conference, pp. 63–72. IEEE (2005)
31. Goodwin, A., Wheat, H.: Human tactile discrimination of curvature when contact area with the skin remains constant. Exp. Brain Res. **88**, 447–450 (1992)
32. Goodwin, A., Browning, A., Wheat, H.: Representation of curved surfaces in responses of mechanoreceptive afferent fibers innervating the monkey's fingerpad. J. Neurosci. **15**(1), 798–810 (1995)
33. Goodwin, A., Macefield, V., Bisley, J.: Encoding of object curvature by tactile afferents from human fingers. J. Neurophysiol. **78**(6), 2881–2888 (1997)
34. Gordon, I.E., Morison, V.: The haptic perception of curvature. Percept. Psychophys. **31**(5), 446–450 (1982)
35. Gordon, I., Voos, A.C., Bennett, R.H., Bolling, D.Z., Pelphrey, K.A., Kaiser, M.D.: Brain mechanisms for processing affective touch. Human Brain Mapping **34**(4), 914–922 (2013)
36. Greenspan, J.D., Bolanowski, S.J.: The psychophysics of tactile perception and its peripheral physiological basis. In: Pain and Touch, pp. 25–103. Elsevier, Amsterdam (1996)
37. Grigorii, R.V., Colgate, J.E., Klatzky, R.: The spatial profile of skin indentation shapes tactile perception across stimulus frequencies. Sci. Rep. **12**(1), 13185 (2022)
38. Gueorguiev, D., Vezzoli, E., Mouraux, A., Lemaire-Semail, B., Thonnard, J.L.: The tactile perception of transient changes in friction. J. R. Soc. Interface **14**(137), 20170641 (2017)
39. Hauser, S.C., Gerling, G.J.: Force-rate cues reduce object deformation necessary to discriminate compliances harder than the skin. IEEE Trans. Hap. **11**(2), 232–240 (2017)
40. Hauser, S.C., Gerling, G.J.: Imaging the 3-d deformation of the finger pad when interacting with compliant materials. In: 2018 IEEE Haptics Symposium (HAPTICS), pp. 7–13. IEEE (2018)
41. Hauser, S.C., McIntyre, S., Israr, A., Olausson, H., Gerling, G.J.: Uncovering human-to-human physical interactions that underlie emotional and affective touch communication. In: 2019 IEEE World Haptics Conference (WHC), pp. 407–412. IEEE (2019)
42. Hay, E., Pruszynski, J.A.: Orientation processing by synaptic integration across first-order tactile neurons. PLoS Comput. Biol. **16**(12), e1008303 (2020)
43. Duprez, G.H.C., Delhaye, B.P., Delannay, L.: Collagen induces anisotropy in fingertip subcutaneous tissues during contact. IEEE Trans. Haptics **17**(4), 753–760 (2024)
44. Hertenstein, M.J., Keltner, D., App, B., Bulleit, B.A., Jaskolka, A.R.: Touch communicates distinct emotions. Emotion **6**(3), 528 (2006)
45. Hertenstein, M.J., Holmes, R., McCullough, M., Keltner, D.: The communication of emotion via touch. Emotion **9**(4), 566 (2009)
46. Hirota, K., Ujitoko, Y., Sakurai, S., Nojima, T.: Deformation matching: Force computation based on deformation optimization. IEEE Trans. Hap. **15**(2), 267–279 (2022)
47. Hollins, M., Bensmaïa, S.J.: The coding of roughness. Can. J. Exp. Psychol. (Revue canadienne de psychologie expérimentale) **61**(3), 184 (2007)
48. Hollins, M., Bensmaıa, S., Roy, E.: Vibrotaction and texture perception. Behavioural Brain Research **135**(1-2), 51–56 (2002)
49. Hollins, M., Lorenz, F., Harper, D.: Somatosensory coding of roughness: the effect of texture adaptation in direct and indirect touch. J. Neurosci. **26**(20), 5582–5588 (2006)

50. Howe, R.D., Cutkosky, M.R.: Dynamic tactile sensing: Perception of fine surface features with stress rate sensing. IEEE Trans. Robot. Automation **9**(2), 140–151 (1993)
51. Hsiao, S.: Central mechanisms of tactile shape perception. Current Opinion Neurobiol. **18**(4), 418–424 (2008)
52. Huloux, N., Willemet, L., Wiertlewski, M.: How to measure the area of real contact of skin on glass. IEEE Trans. Hap. **14**(2), 235–241 (2021)
53. Hussain, S.H., Limthongkul, B., Humphreys, T.R.: The biomechanical properties of the skin. Dermatologic Surgery **39**(2), 193–203 (2013)
54. Janko, M., Wiertlewski, M., Visell, Y.: Contact geometry and mechanics predict friction forces during tactile surface exploration. Sci. Rep. **8**(1), 4868 (2018)
55. Jenmalm, P., Birznieks, I., Goodwin, A.W., Johansson, R.S.: Influence of object shape on responses of human tactile afferents under conditions characteristic of manipulation. Eur. J. Neurosci. **18**(1), 164–176 (2003)
56. Johansson, R.S., Birznieks, I.: First spikes in ensembles of human tactile afferents code complex spatial fingertip events. Nature Neurosci. **7**(2), 170–177 (2004)
57. Johansson, R.S., Vallbo, A.B.: Tactile sensibility in the human hand: relative and absolute densities of four types of mechanoreceptive units in glabrous skin. J. Physiol. **286**(1), 283–300 (1979)
58. Johnson, K.O., Hsiao, S.S.: Neural mechanisms of tactual form and texture perception. Ann. Rev. Neurosci. **15**, 227–250 (1992)
59. Kappers, A.M.: Human perception of shape from touch. Philos. Trans. R. Soc. B: Biol. Sci. **366**(1581), 3106–3114 (2011)
60. Khalsa, P.S., Friedman, R.M., Srinivasan, M.A., Lamotte, R.H.: Encoding of shape and orientation of objects indented into the monkey fingerpad by populations of slowly and rapidly adapting mechanoreceptors. J. Neurophysiol. **79**(6), 3238–3251 (1998)
61. Kim, S.S., Gomez-Ramirez, M., Thakur, P.H., Hsiao, S.S.: Multimodal interactions between proprioceptive and cutaneous signals in primary somatosensory cortex. Neuron **86**(2), 555–566 (2015)
62. Klatzky, R.L., Lederman, S.J.: Tactile roughness perception with a rigid link interposed between skin and surface. Percept. Psychophys. **61**(4), 591–607 (1999)
63. Klatzky, R.L., Lederman, S.J., Hamilton, C., Grindley, M., Swendsen, R.H.: Feeling textures through a probe: Effects of probe and surface geometry and exploratory factors. Percept. Psychophys. **65**, 613–631 (2003)
64. Kwiatkowska, M., Franklin, S., Hendriks, C., Kwiatkowski, K.: Friction and deformation behaviour of human skin. Wear **267**(5-8), 1264–1273 (2009)
65. LaMotte, R.H., Srinivasan, M.A.: Tactile discrimination of shape: responses of rapidly adapting mechanoreceptive afferents to a step stroked across the monkey fingerpad. J. Neurosci. **7**(6), 1672–1681 (1987)
66. LaMotte, R.H., Srinivasan, M.A.: Responses of cutaneous mechanoreceptors to the shape of objects applied to the primate fingerpad. Acta Psychologica **84**(1), 41–51 (1993)
67. LaMotte, R.H., Srinivasan, M.A.: Neural encoding of shape: responses of cutaneous mechanoreceptors to a wavy surface stroked across the monkey fingerpad. J. Neurophysiol. **76**(6), 3787–3797 (1996)
68. Lederman, S.J.: Tactile roughness of grooved surfaces: The touching process and effects of macro-and microsurface structure. Perception & Psychophysics **16**(2), 385–395 (1974)
69. Lederman, S.J.: Tactual roughness perception: spatial and temporal determinants. Can. J. Psychol. (Revue Canadienne de Psychologie) **37**(4), 498 (1983)
70. Li, B., Gerling, G.J.: Individual differences impacting skin deformation and tactile discrimination with compliant elastic surfaces. In: 2021 IEEE World Haptics Conference (WHC), pp. 721–726. IEEE (2021)
71. Li, B., Hauser, S., Gerling, G.J.: Identifying 3-d spatiotemporal skin deformation cues evoked in interacting with compliant elastic surfaces. In: 2020 IEEE Haptics Symposium (HAPTICS), pp. 35–40. IEEE (2020)

72. Li, B., Hauser, S.C., Gerling, G.J.: Faster indentation influences skin deformation to reduce tactile discriminability of compliant objects. IEEE Trans. Hap. **16**(2), 215–227 (2023)
73. Lim, S.C., Lee, H.K., Park, J.: Role of combined tactile and kinesthetic feedback in minimally invasive surgery. Int. J. Med. Robot. Comput. Assisted Surgery **11**(3), 360–374 (2015)
74. Macefield, V.G., Häger-Ross, C., Johansson, R.S.: Control of grip force during restraint of an object held between finger and thumb: responses of cutaneous afferents from the digits. Exp. Brain Res. **108**, 155–171 (1996)
75. Manita, S., Suzuki, T., Homma, C., Matsumoto, T., Odagawa, M., Yamada, K., Ota, K., Matsubara, C., Inutsuka, A., Sato, M., et al.: A top-down cortical circuit for accurate sensory perception. Neuron **86**(5), 1304–1316 (2015)
76. Mano, T., Iwase, S., Toma, S.: Microneurography as a tool in clinical neurophysiology to investigate peripheral neural traffic in humans. Clin. Neurophysiol. **117**(11), 2357–2384 (2006)
77. Masen, M.A.: A systems based experimental approach to tactile friction. J. Mech. Behavior Biomed. Mater. **4**(8), 1620–1626 (2011)
78. McGlone, F., Wessberg, J., Olausson, H.: Discriminative and affective touch: sensing and feeling. Neuron **82**(4), 737–755 (2014)
79. Meyer, D.J., Peshkin, M.A., Colgate, J.E.: Fingertip friction modulation due to electrostatic attraction. In: 2013 World Haptics Conference (WHC), pp. 43–48. IEEE (2013)
80. Miguel, E., D'Angelo, M.L., Cannella, F., Bianchi, M., Memeo, M., Bicchi, A., Caldwell, D.G., Otaduy, M.A.: Characterization of nonlinear finger pad mechanics for tactile rendering. In: 2015 IEEE World Haptics Conference (WHC), pp. 63–68. IEEE (2015)
81. Moscatelli, A., Bianchi, M., Serio, A., Terekhov, A., Hayward, V., Ernst, M.O., Bicchi, A.: The change in fingertip contact area as a novel proprioceptive cue. Current Biol. **26**(9), 1159–1163 (2016)
82. Nakatani, M., Sato, A., Tachi, S., Hayward, V.: Tactile illusion caused by tangential skin strain and analysis in terms of skin deformation. In: Haptics: Perception, Devices and Scenarios: 6th International Conference, EuroHaptics 2008 Madrid, Spain, June 10–13, 2008 Proceedings 6, pp. 229–237. Springer (2008)
83. Nicolelis, M.A., Ghazanfar, A.A., Stambaugh, C.R., Oliveira, L.M., Laubach, M., Chapin, J.K., Nelson, R.J., Kaas, J.H.: Simultaneous encoding of tactile information by three primate cortical areas. Nature Neurosci. **1**(7), 621–630 (1998)
84. Pacchierotti, C., Chinello, F., Malvezzi, M., Meli, L., Prattichizzo, D.: Two finger grasping simulation with cutaneous and kinesthetic force feedback. In: Haptics: Perception, Devices, Mobility, and Communication: International Conference, EuroHaptics 2012, Tampere, Finland, June 13–15, 2012. Proceedings, Part I, pp. 373–382. Springer (2012)
85. Phillips, J.R., Johnson, K.O.: Tactile spatial resolution. iii. a continuum mechanics model of skin predicting mechanoreceptor responses to bars, edges, and gratings. J. Neurophysiol. **46**(6), 1204–1225 (1981)
86. Pont, S.C., Kappers, A.M., Koenderink, J.J.: Haptic curvature discrimination at several regions of the hand. Percept. Psychophys. **59**, 1225–1240 (1997)
87. Pont, S.C., Kappers, A.M., Koenderink, J.J.: Similar mechanisms underlie curvature comparison by static and dynamic touch. Percept. Psychophys. **61**, 874–894 (1999)
88. Provancher, W.R., Sylvester, N.D.: Fingerpad skin stretch increases the perception of virtual friction. IEEE Trans. Hap. **2**(4), 212–223 (2009)
89. Pruszynski, J.A., Johansson, R.S.: Edge-orientation processing in first-order tactile neurons. Nature Neurosci. **17**(10), 1404–1409 (2014)
90. Quek, Z.F., Schorr, S.B., Nisky, I., Okamura, A.M., Provancher, W.R.: Augmentation of stiffness perception with a 1-degree-of-freedom skin stretch device. IEEE Trans. Human-Mach. Syst. **44**(6), 731–742 (2014)
91. Rauss, K.S., Pourtois, G., Vuilleumier, P., Schwartz, S.: Attentional load modifies early activity in human primary visual cortex. Human Brain Mapping **30**(5), 1723–1733 (2009)
92. Sanders, A.F., Kappers, A.M.: Bimanual curvature discrimination of hand-sized surfaces placed at different positions. Percept. Psychophys. **68**(7), 1094–1106 (2006)

93. Sato, T., Okada, Y., Miyamoto, T., Fujiyama, R.: Distributions of sensory spots in the hand and two-point discrimination thresholds in the hand, face and mouth in dental students. J. Physiol.-Paris **93**(3), 245–250 (1999)
94. Schirmer, A., Gunter, T.C.: The right touch: Stroking of ct-innervated skin promotes vocal emotion processing. Cogn. Affective Behavioral Neurosci. **17**, 1129–1140 (2017)
95. Shoemaker, J.K., Klassen, S.A., Badrov, M.B., Fadel, P.J.: Fifty years of microneurography: learning the language of the peripheral sympathetic nervous system in humans. J. Neurophysiol. **119**(5), 1731–1744 (2018)
96. Smith, A.M., Chapman, C.E., Deslandes, M., Langlais, J.S., Thibodeau, M.P.: Role of friction and tangential force variation in the subjective scaling of tactile roughness. Exp. Brain Res. **144**, 211–223 (2002)
97. Srinivasan, M.A., LaMotte, R.H.: Tactile discrimination of shape: responses of slowly and rapidly adapting mechanoreceptive afferents to a step indented into the monkey fingerpad. J. Neurosci. **7**(6), 1682–1697 (1987)
98. Srinivasan, M.A., Whitehouse, J., LaMotte, R.H.: Tactile detection of slip: surface microgeometry and peripheral neural codes. J. Neurophysiol. **63**(6), 1323–1332 (1990)
99. Sripati, A.P., Bensmaia, S.J., Johnson, K.O.: A continuum mechanical model of mechanoreceptive afferent responses to indented spatial patterns. J. Neurophysiol. **95**(6), 3852–3864 (2006)
100. Sukumar, V., Johansson, R.S., Pruszynski, J.A.: Precise and stable edge orientation signaling by human first-order tactile neurons. eLife **11**, e81476 (2022)
101. Thompson, E.H., Hampton, J.A.: The effect of relationship status on communicating emotions through touch. Cogn. Emotion **25**(2), 295–306 (2011)
102. Tong, J., Mao, O., Goldreich, D.: Two-point orientation discrimination versus the traditional two-point test for tactile spatial acuity assessment. Front. Human Neurosci. **7**, 579 (2013)
103. Trojahn, C., Dobos, G., Schario, M., Ludriksone, L., Blume-Peytavi, U., Kottner, J.: Relation between skin micro-topography, roughness, and skin age. Skin Res. Technol. **21**(1), 69–75 (2015)
104. Vallbo, A.B., Hagbarth, K.E., Wallin, B.G.: Microneurography: how the technique developed and its role in the investigation of the sympathetic nervous system. J. Appl. Physiol. **96**(4), 1262–1269 (2004)
105. Van Boven, R.W., Johnson, K.O.: The limit of tactile spatial resolution in humans: grating orientation discrimination at the lip, tongue, and finger. Neurology **44**(12), 2361–2361 (1994)
106. Voisin, J., Lamarre, Y., Chapman, C.E.: Haptic discrimination of object shape in humans: contribution of cutaneous and proprioceptive inputs. Exp. Brain Res. **145**, 251–260 (2002)
107. von Mohr, M., Crowley, M.J., Walthall, J., Mayes, L.C., Pelphrey, K.A., Rutherford, H.J.: Eeg captures affective touch: Ct-optimal touch and neural oscillations. Cogn. Affective Behavioral Neurosci. **18**, 155–166 (2018)
108. Watkins, R.H., Dione, M., Ackerley, R., Backlund Wasling, H., Wessberg, J., Löken, L.S.: Evidence for sparse c-tactile afferent innervation of glabrous human hand skin. J. Neurophysiol. **125**(1), 232–237 (2021)
109. Westling, G., Johansson, R.S.: Responses in glabrous skin mechanoreceptors during precision grip in humans. Exp. Brain Res. **66**, 128–140 (1987)
110. Wijntjes, M.W., Sato, A., Hayward, V., Kappers, A.M.: Local surface orientation dominates haptic curvature discrimination. IEEE Trans. Hap. **2**(2), 94–102 (2009)
111. Witney, A.G., Wing, A., Thonnard, J.L., Smith, A.M.: The cutaneous contribution to adaptive precision grip. Trends Neurosci. **27**(10), 637–643 (2004)
112. Won, S.Y., Kim, H.K., Kim, M.E., Kim, K.S.: Two-point discrimination values vary depending on test site, sex and test modality in the orofacial region: a preliminary study. J. Appl. Oral Sci. **25**, 427–435 (2017)
113. Xu, C., Hauser, S.C., Wang, Y., Gerling, G.J.: Roles of force cues and proprioceptive joint angles in active exploration of compliant objects. In: 2019 IEEE World Haptics Conference (WHC), pp. 353–358. IEEE (2019)

114. Xu, S., Xu, C., McIntyre, S., Olausson, H., Gerling, G.J.: Subtle contact nuances in the delivery of human-to-human touch distinguish emotional sentiment. IEEE Trans. Hap. **15**(1), 97–102 (2021)
115. Xu, S., Xu, C., Gerling, G.J.: Social and emotional touch between romantic partners is affectively more pleasant due to finely tuned contact interactions. IEEE Trans. Hap. **16**(3), 449–454 (2023)
116. Yau, J.M., Pasupathy, A., Fitzgerald, P.J., Hsiao, S.S., Connor, C.E.: Analogous intermediate shape coding in vision and touch. Proc. Natl. Acad. Sci. **106**(38), 16457–16462 (2009)
117. Yau, J.M., Connor, C.E., Hsiao, S.S.: Representation of tactile curvature in macaque somatosensory area 2. J. Neurophysiol. **109**(12), 2999–3012 (2013)
118. Yau, J.M., Kim, S.S., Thakur, P.H., Bensmaia, S.J.: Feeling form: the neural basis of haptic shape perception. J. Neurophysiol. **115**(2), 631–642 (2016)
119. Yoshioka, T., Gibb, B., Dorsch, A.K., Hsiao, S.S., Johnson, K.O.: Neural coding mechanisms underlying perceived roughness of finely textured surfaces. J. Neurosci. **21**(17), 6905–6916 (2001)

Chapter 3
Second Foundational Principle: Individual Differences in Perception

Abstract Touch, though governed by universal physical principles, is experienced idiosyncratically. The first foundation established how skin mechanics and psychophysics define the pathways by which tactile information is gathered, transmitted, and encoded—forming the baseline operating envelope for haptic systems. The present chapter extends that foundation by addressing the variability of human touch: how anatomical, biomechanical, and cognitive differences shape what individuals feel, and why this variability must be explicitly accounted for in the design of integrated haptic systems.

3.1 Variability in Human Touch

Tactile experience differs across people. A surgeon distinguishing tissue stiffness, a child squeezing a toy, and an older adult pressing the same object each encounter distinct perceptual realities despite identical physical stimuli. These variations arise from measurable differences in skin stiffness, hydration, thickness, mechanoreceptor density, and neural processing. Psychophysical and biomechanical studies have shown systematic effects of demographic factors (age, gender), anatomical traits (finger size, ridge density), and physiological conditions (hydration, temperature, fatigue) on tactile sensitivity and discrimination thresholds.

From a scientific standpoint, these differences define the boundaries of the human tactile system: they reveal how the same physical deformation can produce distinct perceptual outcomes depending on the observer. From an engineering standpoint, they highlight why haptic systems must be designed not for an "average" user but for a population of users with varying perceptual envelopes. Systems that adapt to these differences by sensing, calibrating, or compensating for user-specific thresholds achieve more reliable and intuitive haptic performance.

This dual perspective constitutes the second foundation for integrated haptic system design: understanding and accommodating individual differences in tactile perception. While the first foundation explained how tactile information is mechanically encoded and psychophysically transformed, the second ensures that these processes remain stable and meaningful across users and contexts. Insights from

B. Li, T. Cody, *Design Principles for Integrated Haptic Systems*, Springer Series on Touch and Haptic Systems, https://doi.org/10.1007/978-3-032-16986-0_3

this foundation guide design for hardware, software, and application subsystems to ensure consistent and perceptually valid performance within diverse human populations.

Foundational Principle 2—Individual Variability:
Tactile perception varies systematically across individuals due to both physical (mechanical) and informational (neural–cognitive) differences arising in skin mechanics, neural encoding, and perceptual interpretation. These variations manifest as measurable shifts in psychophysical response and sensitivity to mechanical cues.

Compliance perception will serve as a running case study throughout the remainder of the book. It illustrates how fundamental variability in skin mechanics translates into perceptual differences, and how these differences can be addressed through an integrated haptic system design process. By following compliance perception from foundational scientific studies that reveal individual differences, through to the translation of these findings into subsystem design goals for hardware, software, and application subsystems to form integrated haptic systems such as artificial skin, this book demonstrates how an integrated, inference-centric haptic system can realize compliance perception as a consistent, perceptually meaningful haptic capability within a broader system across users.

This chapter explores the scientific basis of individual differences in tactile perception, beginning with findings from the literature on how anatomical and biomechanical factors—such as finger size, fingerprint ridge patterns, and skin properties including stiffness, thickness, conformance, and hydration—affect the way tactile information is sensed and interpreted. Then, the case study of compliance perception will be presented, drawing on biomechanical and psychophysical experiments that empirically demonstrate how these factors, particularly skin stiffness, influence perceptual outcomes. The chapter concludes by discussing the implications of these findings for design, showing how an understanding of individual differences can inform subsystem goals and architecture.

3.2 Role of Finger Size and Fingerprint Ridges on Tactile Perception

3.2.1 Finger Size

Finger size is a crucial factor in tactile perception because it determines the density of mechanoreceptors—particularly Merkel cells responsible for fine spatial resolution. The density of these mechanoreceptors is not uniform across individuals and is significantly influenced by finger size [2, 6, 32, 33]. Studies have shown that the surface area of the fingertip, whether in a static or dynamic state, directly influences the number of receptors per unit area. Since a smaller fingertip packs these receptors more densely, individuals with small fingers tend to have superior tactile acuity compared to those with larger fingers due to a higher resolution for tactile information.

The impact of finger size on tactile acuity has been demonstrated across different age groups and between genders. For instance, findings indicate that during childhood, as the fingertips grow, there is a gradual decline in innervation density due to the expanding skin surface, leading to larger receptive fields and reduced spatial resolution. However, this potential decline is partly compensated by maturation in the central nervous system, maintaining relatively stable tactile acuity across childhood [32]. Similarly, studies have shown that females generally exhibit better tactile performance than male because they have smaller fingers in general [3]. These findings highlight the dynamic interplay between peripheral and central mechanisms in shaping tactile perception.

Beyond mechanoreceptor density, finger size also affects real contact area and touch gestures, further influencing tactile perception [2]. Larger fingers have a broader contact area, which alters the mechanical interaction with surfaces and the way forces are distributed across the skin. This can affect the engagement of mechanoreceptors and the anisotropic properties of physical contact, meaning that tactile perception can vary depending on the direction and manner in which an object is explored.

3.2.2 *Finger Ridge Density, Breadth, and Patterns*

In addition to finger size, fingerprint ridges play a crucial role in shaping tactile perception by structuring the spatial organization and sensitivity of mechanoreceptors in the skin [7, 21, 36]. Studies demonstrate that the response of first-order tactile neurons, specifically fast-adapting (RA-I) and slowly-adapting (SA-I) afferents, which encode fine spatial and temporal details of touch, have receptive fields that align with individual fingerprint ridges [19]. These neurons contain multiple highly sensitive zones, or "subfields", that correspond to the spacing of papillary ridges, ensuring that tactile stimuli are processed with high spatial resolution. The spatial sensitivity of these subfields is around 0.4 mm, which closely matches the typical spacing of fingerprint ridges. This alignment ensures that mechanoreceptors within a ridge function as independent units for tactile input, enabling high spatial resolution. One of the key findings from the research is that the spatial arrangement of these receptive subfields remains stable across different scanning speeds. Regardless of whether the fingertip moves slowly or quickly over a surface, mechanoreceptor activation remains finely tuned to the ridge structure, suggesting that fingerprint ridges function as a biomechanical filter. This filtering effect helps maintain a reliable neural representation of tactile information under dynamic conditions, ensuring consistent perception and recognition. While the spatial organization of mechanoreceptors remains constant, scanning direction can slightly shift subfield activation due to shear deformations of the skin [19]. This shift suggests that fingerprint ridges contribute to directional adaptation, allowing for enhanced sensitivity to surface features that exhibit anisotropic properties. By permitting subtle displacements in mechanoreceptor activation, ridges help optimize the detection of fine details across different movement directions, improving tactile sensitivity for complex textures and structures.

Finger ridge characteristics, such as ridge density, ridge breadth, and overall patterns, can shape the way tactile information is encoded and processed. These dermatoglyphic features influence the density and spatial organization of mechanoreceptors, particularly Meissner corpuscles and Merkel cells, thereby affecting both neural coding and skin mechanics in tactile interactions [33].

Ridge density, which refers to the number of ridges per unit area, is observed to vary by gender, age, and population and plays a key role in tactile acuity. Research has found that ridge density decreases with age, likely due to changes in skin elasticity and ridge breadth expansion [16]. Notably, females generally exhibit higher ridge densities than males after puberty, aligning with their typically superior tactile acuity. Higher ridge densities likely support greater mechanoreceptor density per unit area which improves spatial resolution and touch sensitivity. Ridge breadth, or the width of individual ridges, on the other hand, shows inverse correlation with mechanoreceptor density. Studies show that individuals with broader ridges tend to have a lower density of Meissner corpuscles, as these mechanoreceptors are typically located within ridge structures [10]. Since males generally have larger fingers with broader ridges, they exhibit lower receptor destiny per unit length, potentially contributing to their reduced tactile sensitivity compared to females. However, despite this lower receptor density, males and females tend to have a similar absolute number of Meissner corpuscles per finger, suggesting that receptor spatial distribution, rather than total receptor count, drives differences in tactile acuity.

Moreover, dermatoglyphic patterns, including loops, whorls, and arches, also influence tactile perception. The minute details of ridge patterns, such as the number of junctions within a fingerprint, may serve as biomechanical anchors that alter the deformation of the skin during contact. Research indicates that the type of these patterns, along with their specific minutiae, can modulate how tactile stimuli are processed. For instance, a medium-sized loop pattern with a greater number of junctions (particularly in males), has been associated with optimal tactile performance, possibly due to a more efficient organization of the underlying mechanoreceptive fields [27, 28]. This intricate arrangement appears to facilitate the neural coding process, enabling the nervous system to capture and process fine tactile details more effectively.

Beyond simply housing mechanoreceptors, the unique microstructure of fingerprint ridges actively enhances tactile sensitivity by modifying mechanical signal transmission. For example, a biomimetic sensor is used to demonstrate that fingerprint ridges act as a spectral filter for tactile information, amplifying vibrations within the frequency range most efficiently processed by Pacinian corpuscles

[37]. Similarly, studies have been conducted to examine how intermediate ridges beneath the papillary ridges contribute to mechanoreceptor function [13, 34]. The microstructure of the fingerprint, specifically the interplay between the dense papillary ridges and the opposing intermediate ridges, further refines tactile perception. The papillary ridges, which house sweat ducts and sites of afferent innervation, serve as critical zones for initiating tactile signals. In contrast, the underlying intermediate ridges contribute to the mechanical stiffness of the skin layers [13]. While early theories proposed that papillary ridges directly amplify mechanoreceptor response through a lever mechanism, finite element models suggest a more complex interaction. These models suggest that the intermediate ridges can amplify the mechanical input received at the papillary ridges, particularly enhancing the sensitivity of SA-I fibers. Moreover, the changes in the stiffness of these skin layers can significantly modulate the neural response to tactile stimuli, thus fine-tuning our sense of touch. The 3D organization of skin layers further refines tactile perception by modulating strain energy distribution at mechanoreceptor sites. Simulation models reveal that the 3D structure of dermal papillae has a focalization effect on strain energy density, concentrating mechanical stimuli at specific receptor locations [34]. This focalization enhances local sensitivity, particularly in RA-I afferents (Meissner corpuscles), improving their response to fine texture and motion cues, as well as facilitate edge discrimination by detecting subtle changes in surface topography with greater precision.

Finger size and finger ridge characteristics are not merely passive anatomical features: they actively shape our sense of touch by influencing both the density and arrangement of mechanoreceptors, and thus the neural coding of tactile information. Smaller fingers tend to concentrate receptors more densely (e.g., Merkel cells beneath the sweat pores), enhancing spatial resolution, while the precise microstructure of fingerprint ridges organizes receptor subfields that capture mechanical events at the scale of individual ridges. Together, these factors create a detailed peripheral sensory input that is essential for the high fidelity of tactile perception. To bridge the gap between these peripheral inputs and central processing outcomes, it is also important to consider the properties of the skin itself, which further modulate the tactile signals transmitted to the brain.

3.3 Skin's Properties and Their Effect on Shaping Tactile Perception

Skin properties—such as stiffness, thickness, and hydration—are influential to tactile discriminative tasks, as they affect how mechanical events are transmitted and processed by the underlying mechanoreceptors. While anatomical features like finger size and fingerprint ridges provide structural constraints for sensory encoding, the skin's mechanical properties actively modulate how external stimuli deform the skin surface and stimulate tactile afferents. Variability in skin stiffness can lead to differences in contact mechanics, affecting how mechanoreceptors detect physical property of an object. Similarly, skin thickness determines the depth at which mechanical forces propagate, altering the spatial resolution of touch. Hydration levels influence friction at skin-surface interface and the skin's ability to conform to surfaces. Because these properties change due to age, environmental conditions, and prior experiences, they are critical factors in understanding tactile perception and its variability across individuals, which can be used to inform the design of advanced haptic systems that effectively mimic human touch capabilities and are adaptive to diverse users and applications.

3.3.1 Skin Stiffness

Skin stiffness plays a critical role in shaping tactile perception because it fundamentally modulates how tactile information is transduced and perceived. Mechanically, stiffness defines the relationship between applied force and resulting deformation, thereby controlling the magnitude, distribution, and temporal dynamics of strain fields in the skin during contact. Softer, more compliant skin exhibits greater deformation under the same load, resulting in larger contact areas and more pronounced gradients of strain and curvature across the surface. Conversely, stiffer skin resists deformation, producing smaller, more localized contact regions and weaker spatial gradients. These differences directly influence the physical cues available to the mechanoreceptors embedded within the skin. Consequently, even under identical contact conditions, two individuals with different skin stiffness profiles may generate distinct tactile cues at the skin-surface interface, leading to divergent perceptual experiences.

At the level of skin mechanics, stiffer skin exhibits shallower indentation and smaller contact area for a given force, which limits the magnitude and diversity of cues available to signal object properties such as compliance, texture, and surface curvature [8, 23]. In particular, the spatial extent and amplitude of strain modulations affect the activation patterns of slowly adapting type I (SA-I) and rapidly adapting (RA-I) afferents, which encode local pressure and dynamic deformation [35, 38]. In softer skin, richer deformation cues enable finer resolution of rendered properties, whereas stiffer skin filters out subtle variations in deformation, leading to attenuated receptor responses and reduced discriminability in psychophysical tasks. These effects are supported by both finite element simulations and experimental findings, which show that skin stiffness significantly alters the mechanical transduction of tactile stimuli and correlates with psychophysical performance in discrimination tasks [12, 23, 24].

At the level of neural processing, variations in stiffness alter both the encoding and interpretation of tactile input. Mechanoreceptor responses form the basis for higher-order neural representations of tactile attributes within the somatosensory cortex [14, 30]. When mechanical cues are weakened or less distinct due to high skin stiffness, cortical neurons receive less differentiated input, resulting in lower perceptual contrast between stimuli. Moreover, stiffer skin can reduce temporal variation in afferent firing patterns and spatial recruitment, diminishing the information available for tactile percepts. This modulation in peripheral input propagates to central somatosensory processing, altering signal-to-noise characteristics and potentially impairing perceptual accuracy. While the human nervous system may adapt to such variability through experience or active control, most haptic devices lack the capacity to account for or adapt to user-specific skin mechanics, resulting in inconsistent or non-intuitive feedback.

3.3.2 Skin Conformance

Closely correlated with skin stiffness, skin conformance—the ability of the skin to conform to surfaces—directly affects contact mechanics and neural activity in the afferent fibers, shaping how tactile information is processed. Studies demonstrate that more compliant skin deforms more easily, allowing it to conform more closely to fine spatial detail, which leads to better performance in spatial discrimination tasks such as grating orientation and gap detection [15, 39]. The increased conformance enhances the mechanical contrast at the level of mechanoreceptors, particularly SA-I afferents, which are responsible for detecting fine spatial details. When the skin conforms well, the spatial resolution of the tactile input is enhanced, leading to lower discrimination thresholds and finer perceptual resolution.

Additionally, skin conformance is found to be higher along than across the fingertip ridges, which leads to better perceptual discrimination in grating orientation tasks when the gratings are aligned with the ridges. This suggests that the directional anisotropy of skin influences how tactile stimuli are encoded, potentially optimizing perception in certain orientations.

3.3.3 Skin Thickness

Skin thickness has impact on tactile perception by influencing how mechanical forces and vibratory signals are transmitted to mechanoreceptors. The multilayered structure, including the epidermis, dermis, and subcutaneous tissue, contributes to the overall mechanical properties of the skin. Variations in skin thickness affect mechanoreceptor activation, force transmission, and vibratory signal propagation, which in turn modulate the ability to perceive tactile information. Finite element models and experimental measurements reveal that changes in skin thickness impact the degree of skin deformation during contact [4]. In younger skin, optimal thickness and elasticity enable the skin to conform closely to an object's surface, thereby enhancing the mechanical signals that mechanoreceptors rely on for high-resolution tactile discrimination. In contrast, with aging, the skin becomes thinner due to reductions in epidermal and dermal layer thickness, as well as the degradation of structural proteins like collagen and elastin. This altered thickness and diminished mechanical integrity of the skin result in a reduced ability to detect these subtle cues.

The thickness of the skin also affects how mechanical forces propagate through the tissue and reach mechanoreceptors. Thicker skin acts as a mechanical buffer, altering the way external forces are distributed. Then the skin is thicker, the applied force is spread over a larger volume of tissue, leading to reduced localized strain on mechanoreceptors. As a result, individuals with thicker skin may experience poorer discrimination because the deformation signals reaching SA-I afferents are less pronounced. Conversely, thinner skin allows mechanical forces to be transmitted more directly, potentially enhancing spatial resolution and tactile sensitivity. This is consistent with the findings which demonstrate that higher skin compliance (often associated with thinner skin) improves tactile acuity in younger individuals [39].

Beyond force transmission, skin thickness also affects vibratory sensitivity, which is critical for texture perception and edge detection. The results from evaluating how skin thickness influences the transmission of vibratory signals show that aging-related skin thinning leads to a decline in vibratory sensitivity, particularly in high-frequency detection. Thicker skin can absorb and dissipate high-frequency vibrations before they reach deeper mechanoreceptors, reducing the effectiveness of RA-II afferents (i.e., Pacinian corpuscles) in encoding micro-texture information. While thinner skin can facilitate better transmission of vibrations to receptors, it may lack the damping effects needed to regulate touch sensitivity. This could lead to increased susceptibility to tactile fatigue or discomfort, especially in environments with repetitive vibratory stimuli.

In summary, skin thickness is a dynamic property that affects tactile perception by modulating skin deformation and how mechanical forces and vibratory signals are transmitted to mechanoreceptors. Thicker skin dampens spatial details and filters out high-frequency vibrations, potentially reducing tactile acuity and discrimination sensitivity. On the other hand, thinner skin improves force transmission and spatial resolution but may increase tactile fatigue. Skin thinning has been observed with aging which may help explain the degradation of tactile acuity from a biomechanical point of view; however, neural deterioration should also be considered for the cause of sensory decline.

3.3.4 Skin Hydration

Skin hydration can significantly modify the mechanical properties of the skin, which in turn affects tactile perception. One of the primary ways in which hydration influences compliance perception is through changes in skin stiffness and elasticity. Hydration softens the stratum corneum—the outermost layer of the skin—making it more pliable and better able to conform to an object's surface. This reduced stiffness and improved conformability facilitate a more precise assessment of the object's mechanical properties, effectively lowering the tactile discrimination threshold. In essence, hydration allows the skin to deform more readily under applied pressure and these subtle deformations can be transmitted as sensory signals to faithfully perceive differences in material compliance. For example, when applying a moisturizer to elderly individuals, their ability to perceive compliance differences increases, likely due to increased skin softness (i.e., decreased stiffness) that increases contact area during tactile contact [22]. This aligned with a study that observes instant improvements in compliance discrimination performance after reducing skin stiffness using hyaluronic acid [24].

Another critical factor influenced by hydration is the transmission of mechanical signals to deeper layers of the skin, where mechanical-sensitive receptors are located. Studies found that hydration modifies the way external forces propagate through the skin, which can either enhance or dampen the perception of compliance [40]. That is, while a softer, more hydrated skin can lead to improved contact and more detailed deformation patterns, enhancing the contrast between soft and hard stimuli, it also tends to dampen the transfer of mechanical energy to the underlying sensory receptors. Microneurographic studies have shown that following hydration, the afferent messages conveyed by tactile neurons are depressed, suggesting that the softer skin reduces the intensity of the mechanical signals reaching the receptors, which might be counterbalanced by the improved spatial resolution due to reduced skin stiffness.

Furthermore, hydration-induced changes are not uniform across all types of tactile perception. For example, while vibrotactile detection thresholds seem relatively unaffected by hydration, the perception of roughness and texture is substantially altered due to its dependency on skin deformation cues. Hydration also modifies the skin's viscoelastic properties, influencing how dynamic mechanical cues are transmitted through skin structures. In compliance discrimination, a hydrated skin surface may produce a more pronounced deformation pattern, such as larger changes in contact area and penetration depth, thereby allowing for more reliable encoding of compliance differences.

These findings underscore the need to consider skin hydration in modeling tactile perception. In fact, the effects of hydration are particularly important in the context of aging. As individuals age, skin hydration decreases naturally, leading to reduced elasticity and impaired afferent function [9]. This decline in hydration is a major contributor to age-related tactile deterioration, as observed in two-point discrimination and texture perception tasks. Therefore, considerations of skin hydration in design goals could potentially improve applications in haptic technology and sensory rehabilitation.

> The physical properties of the skin are not passive layers but active filters that fundamentally shape tactile perception by modulating how mechanical stimuli are transmitted to mechanoreceptors.

(continued)

- **Skin Stiffness:** Softer skin deforms more, creating larger contact areas and richer strain gradients, which enhances the perception of compliance and fine spatial details. Stiffer skin resists deformation, leading to weaker neural signals and reduced discriminability.
- **Skin Conformance:** More compliant skin conforms more closely to surface details, improving performance in spatial tasks like grating orientation and gap detection by enhancing the mechanical contrast for SA-I afferents.
- **Skin Thickness:** Thicker skin acts as a mechanical buffer, dampening spatial details and high-frequency vibrations, which can reduce tactile acuity. Thinner skin allows more direct force transmission, potentially improving spatial resolution.
- **Skin Hydration:** Hydration softens the skin, improving conformance and compliance discrimination. However, it can also dampen the transmission of mechanical energy to deeper receptors, creating a complex trade-off between spatial resolution and signal intensity.

These properties—which vary with age, environment, and individual—act as a unique biomechanical lens for each person, directly influencing the fidelity of the tactile information their nervous system receives.

3.4 Effects of Age and Gender

Human tactile sensitivity and perception are not uniform: spatial acuity, vibration thresholds, and discrimination ability all change systematically across lifespan and differ measurably between males and females. These variations exert profound effects on the mechanical properties of the skin and the neural pathways that transduce touch, making it essential to incorporate these factors into both scientific models and applied engineering designs. Children's fingertips, for example, grow comparatively fast, temporarily widening receptive fields of underlying afferents [32]; while older adults experience progressive reductions in mechanoreceptor density, skin elasticity, and nerve conduction speed—factors that collectively degrade their discriminative touch abilities [4, 5, 9, 31]. Meanwhile, females generally exhibit superior performance in discriminative tasks (e.g., fine-texture, vibration tests) than males due to smaller finger size, higher fingerprint ridge density, and softer skin surface [1, 16, 33]. Accounting for age and gender related variations is imperative for understanding somatosensory processing and a practical necessity for design as it defines the operating envelope within which haptic technologies must function. Any device that seeks to deliver naturalistic feedback—be it a prosthetic hand, a teleoperation controller, or a tactile display in virtual reality—must therefore calibrate the system's actuation amplitudes, sensing bandwidth, and control algorithms to accommodate the full range of age- and gender-dependent sensitivities to be inclusive, reliable, and perceptually faithful.

A variety of experimental paradigms and measurements have been employed to assess how tactile perception is modulated by age and gender. Psychophysical tasks—such as two-point discrimination, compliance differentiation, and grating-orientation thresholds—capture perceptual thresholds across the lifespan. Biomechanical measurements of static and dynamic contact between skin and the object quantify the skin deformation and its viscoelastic and topographical properties. Vibratory testing and frequency analysis decompose the spectral features of touch gestures. Direct interventions are used to reveal causal links between skin properties and tactile sensitivity, while finite element models simulate how changes in skin elasticity and receptor density affect stimulus encoding at the neural level. These methods provide insights into perceptual performance and help uncover the underlying physiological and biomechanical mechanisms that contribute to individual differences. Importantly, the combination of psychophysical testing, biomechanical measurements, and neurophysiological modeling reveal how demographic diversity impacts sense of touch.

One of the foundational psychophysical techniques used to assess spatial acuity is the two-point discrimination task, which measures the minimal separation distance at which two simultaneous contacts are perceived as distinct. Numerous studies using this method have demonstrated a marked decline in tactile acuity with increasing age. For example, studies found that older adults exhibit significantly higher threshold for two-point discrimination than younger individuals due to decreased skin elasticity and reduced peripheral mechanoreceptor density [4, 4, 9, 29]. Specifically, skin becomes less hydrated, less elastic, and structurally thinner as age increases, contributing to a decline in the transmission of mechanical cues. Mechanoreceptors such as Merkel cell and Meissner corpuscles decline in number or function during aging, reducing sensitivity to tactile discriminability.

In the central nervous system, age-related changes include cortical reorganization, synaptic degradation, and decreased somatosensory inhibition. Studies have observed an expansion of hand representation in the primary somatosensory cortex (S1) in older adults, associated with poor tactile acuity, suggesting altered intracortical inhibition that further changes perceptual processing [20, 29]. Additionally, aging leads to general reductions in brain volume, metabolism, and white matter integrity, further impacting somatosensory integration [29]. At the neural level, studies report a decline in large myelinated afferents, reduced conduction velocity, and demyelination in both peripheral and central nervous systems. Notably, not all neural pathways degrade equally—unmyelinated C-tactile afferents, which mediate affective touch responses, appear more resilient to aging. As a result, older adults often report increased perceived pleasantness despite declines in discriminative touch, possibly due to preserved unmyelinated afferents or psychosocial factors like increased desire for interpersonal touch, suggesting a shift in the relative contribution of sensory pathways with age.

In the grating orientation task, a method that effectively studies developmental and anatomical influences on tactile acuity, participants are asked to discern the orientation of ridged gratings pressed against the skin. Studies revealed that individuals, particularly children and adults with smaller fingers, and hence, higher receptor density per unit area, demonstrate superior tactile discrimination [32, 33]. Among young adults, those with more compliant skin show substantially lower detection threshold in discriminating grating orientation than those with stiffer skin, especially when the gratings orient along with the skin [39].

Similar findings have been observed in compliance discrimination tasks, in passive touch, through the use of 3D optical imaging techniques, which provide biomechanical measurements of skin deformation by capturing high-resolution, real-time mappings of the skin's mechanical responses under load [23, 24]. This approach enables a direct examination of how individual differences in skin properties—particularly skin stiffness—relate to tactile perception variations. In the study of evaluating compliance discriminability from participants with varying skin stiffness, the results revealed that individuals with softer skin, more commonly found among females and younger participants, exhibited more pronounced deformation patterns during contact regardless of finger size. Specifically, the statistical results from principal component analysis (PCA) show that these participants (characterized as Group 1 for softer skin) generate higher change rate in skin deformation-based cues including contact area, surface curvature, and penetration depth. In contrast, the participants in Group 2 with stiffer skin, demonstrated higher changes in contact force and contact-shape eccentricity, see Fig. 3.1. These differences in skin deformation pattern influenced by individuals' skin property lead to variations in psychophysical discrimination capability, where participants with softer skin significantly outperformed those with stiffer skin in differentiating both harder and softer stimuli relative to skin compliance. Moreover, such changes in skin's mechanics also influence exploratory strategies and behaviors in active touch. Indeed, certain exploration strategies can improve discriminative performance. For example, studies show that participants with stiff skin tend to apply higher contact forces and larger joint movements to compensate for the limited skin deformation information when discriminating compliant objects [41].

Likewise, studies of contact interface imaging further clarify gender-based differences. Contact surface imaging captures the effective contact patch between the finger and object and is important for estimating ridge density and mechanoreceptor engagement during contact. The results show that females have significantly higher fingerprint ridge density than males, particularly after puberty, which contributes to higher mechanoreceptor density per unit area and better spatial resolution. Moreover, studies further observed that females tend to have higher adhesive contact forces and lower Young's modulus, which amplify skin-surface interactions and enhance perception of compliance and surface texture [1, 16].

Beyond static spatial discrimination, researchers have developed tribohaptic systems and vibratory analysis tools that capture the dynamic interactions between the skin and surface during exploratory touch. These systems are used to analyze how different touch gestures (e.g., lateral versus vertical strokes) and biophysical properties (e.g., skin stiffness, surface roughness) influence perceptual outcomes. A Mel-frequency Cepstral Coefficient (MFCC) pipeline decomposes vibratory signals into spectral features and correlates them with psychophysical ratings. Older adults produce lower-amplitude, lower-frequency signatures that mirror their diminished sensitivity to fine textures, whereas females generate richer, higher-energy spectra and describe more vivid textural impressions, especially when allowed to stroke from left to right, a gesture shown to maximize perceptual gain, likely due to higher skin compliance and greater surface conformity, both of which enhance vibrational transmission and tactile encoding[3, 4]. These findings underscore that touch behavior itself—how individuals explore objects—can mediate perceptual differences rooted in biological variation.

Finite element simulations have been used to assess how age-induced changes in skin structure impact tactile performance. These simulations allow researchers to predict the mechanical and vibrational behavior of aged versus young skin under controlled conditions. For example, multilayered viscoelastic models have been developed for young and aged fingers, and showed that older skin, due to reduced elasticity and increased surface roughness, transmits lower amplitude vibrations and deforms less under the same loading conditions. Similarly, the outcomes of first-order tactile neuron modeling observed that afferent density and skin modulus affect the tactile encoding of stimulus information [4, 9, 17].

Intervention studies have been conducted to directly reflect the causal links between skin mechanics and tactile perception. For instance, experiments using hyaluronic acid to soften participants' fingertips reveal that the softened skin immediately improves perceptual discriminability by 5–30% across all participants, especially for those with stiff skin [24]. Similarly, neurophysiological studies have shown that skin properties influence neural encoding in both the responsiveness of single afferents and the recruitment of populations of afferents. For example, directly modulating the compliance of finger pad through venous occlusion can significantly affect the firing rates of subtypes of cutaneous mechanoreceptors, especially the rapidly adapting afferents [18].

Taken together, these scientific experiments and empirical observations provide an aligned understanding that age and gender modulate tactile sensitivity and perception by changes in skin mechanics that lead to variations in mechanoreceptor distribution, exploratory strategies, and neural coding. Age leads to general declines in tactile discrimination due to peripheral and central degradation, while gender differences emerges from variations in finger size, ridge density, skin properties such as stiffness. Considering the effect of age and gender refines scientific models of somatosensory processing and provides guide towards adaptive haptic designs that tuned for the user's biological profile.

3.5 Individual Differences in Compliance Perception

To empirically demonstrate how anatomical features and skin properties shape tactile perception, a series of biomechanical and psychophysical experiments were conducted to examine compliance discrimination across individuals exhibiting a range of finger sizes, fingerprint ridge patterns, and skin stiffness levels. The perception of compliance, profoundly rooted in how skin deforms under load, offers clear lens through which to examine how individual differences in skin mechanics translate into perceptual differences in tactile events.

Biomechanical Evidence: Distinct Skin Deformation Patterns Recent studies using 3D imaging techniques demonstrated how local skin deformation patterns generate essential cues that contribute to our perception of compliance [25, 26]. By constructing high-resolution 3D point cloud representations of the skin surface at short time intervals during contact with compliant stimuli, skin deformation cues, including contact area, curvature, penetration depth, eccentricity, and contact force, can be effectively observed and analyzed to characterize how the skin responds to different levels of compliance. These cues, which have been previously identified as critical for compliance perception, are shown to vary systematically based on stimulus properties as well as individual difference in skin stiffness, see Fig. 3.1.

The results revealed distinct patterns of skin deformation depending on the compliance of the stimulus. When a stiffer stimulus indents the fingertip, the skin tends to flatten against the substrate, resulting in a larger contact area and greater contact force, accompanied by a more circular (lower eccentricity) contact shape. In contrast, a more compliant stimulus allows the fingertip to retain more of its natural contour and penetrate into the substrate, producing higher surface curvature, deeper penetration, and a more elliptical (higher eccentricity) shape. These deformation patterns corresponded closely to each participant's skin stiffness, finger size, and fingerprint ridge, with skin stiffness shows the highest correlation. These findings suggest that anatomical and skin-mechanical variability lead to distinct deformation profiles, see Fig. 3.2

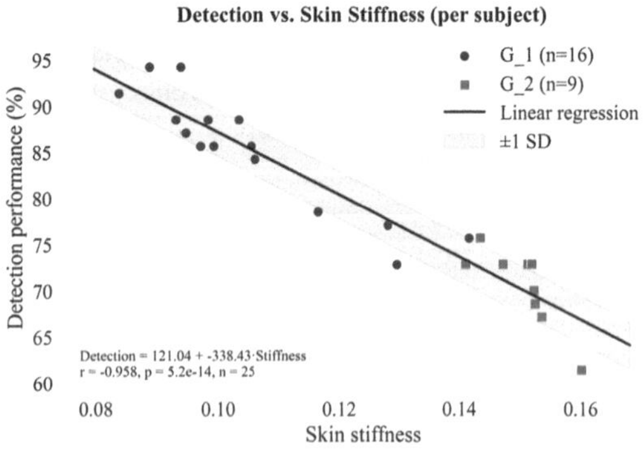

Fig. 3.1 Perceptual sensitivity to compliance is strongly correlated with individual skin stiffness, showing that participants with softer skin achieve higher discrimination performance. G_1 and G_2 correspond to Group 1 and Group 2, respectively

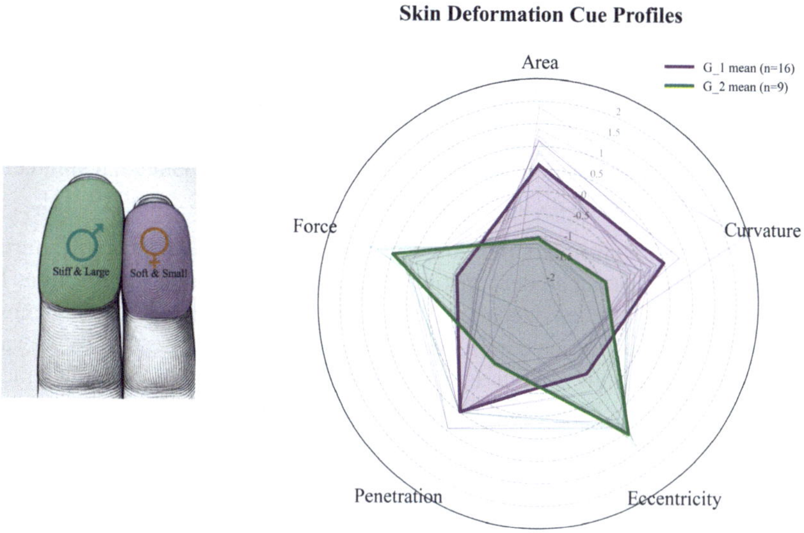

Fig. 3.2 Participants grouped by their skin mechanical properties exhibit distinct deformation cue profiles during contact. Each group relies on different combinations of cues. G_1 corresponds to Group 1 with stiff and large finger pads, and G_2 corresponds to Group 2 with soft and small finger pads. The lighter lines correspond to individual participants

Psychophysical Correlates: Perceptual Consequences of Biomechanical Variability In the study of evaluating compliance discrimination performance, participants were asked to discriminate between pairs of compliant materials in passive touch while both their skin biomechanics and perceptual judgments were recorded [26]. The results reveal that variations in skin mechanics evoke distinct patterns of skin deformation and they are highly correlated with perceptual discriminability, see Fig. 3.3. Specifically, when distinguishing between pairs of compliant stimulus, participants who exhibited larger finger size or stiffer skin showed higher discrimination thresholds. The imaging of skin dynamics during contact indicates that softer

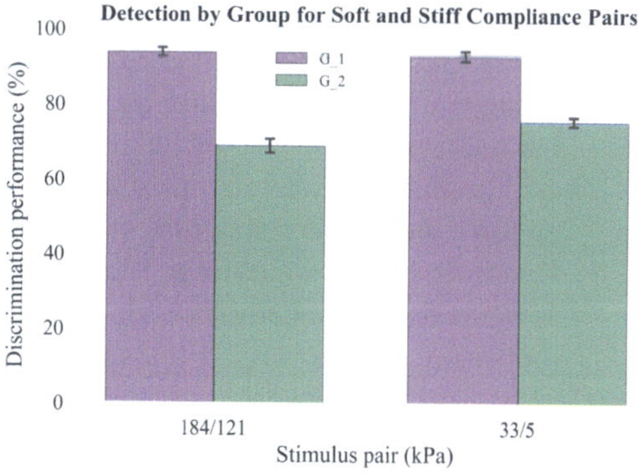

Fig. 3.3 Groups with different skin mechanics show distinct compliance discrimination performance. Participants with smaller, softer fingers (Group 1) outperform those with stiffer fingers (Group 2) for soft and stiff stimuli

fingers generate larger rates of change in contact area and surface curvature between pairs of stimuli, which modulate how the mechanical stimuli propagate to locations of mechanosensitive end organs at the epidermal-dermal border and influence neural firing. Such observations are also supported by scientific evidence in elastography analysis which shows that elastic modulus of the skin affects the distribution of stresses and strains throughout skin layers and in neural computational simulation, where a lower skin elasticity improves the encoding of elastic stimuli [9, 42].

Selection of Deformation Cues for Encoding Compliance Crucially, further analysis revealed that not all deformation cues contribute equally to perception. The relative weighting of cues depended on the stimulus modulus, contact dynamics, individual skin properties, and even demographic factors such as gender. For example, when discriminating stimuli that are softer than the skin, perceptual judgments relied more on the change rate of contact area and surface curvature, whereas for stiffer materials, force rate becomes more informative. These findings indicate that tactile perception is a context-dependent, user-specific integration process, dynamically reweighting available cues based on both the external environment and intrinsic skin mechanics.

Tactile perception is not one-size-fits-all; it is profoundly shaped by an individual's age, gender, and unique biological makeup, which directly influence skin mechanics and neural encoding. These factors create distinct "perceptual profiles" that determine how we interact with and interpret the world through touch.

- **Age-Related Changes:** Aging leads to stiffer, thinner skin and a decline in mechanoreceptor density. This dampens skin deformation, resulting in poorer discrimination of texture, compliance, and spatial details (like in two-point discrimination tests), though the affective touch pathway remains more resilient.
- **Gender Differences:** Females often exhibit superior tactile acuity, driven by smaller finger size, higher fingerprint ridge density, and softer, more compliant skin. This enhances spatial resolution and generates richer skin deformation patterns during contact.
- **Skin Stiffness as a Key Predictor:** Softer skin (more common in the young and females) deforms more, generating larger and more dynamic changes in contact area and curvature. This provides a richer signal to mechanoreceptors, leading to better compliance discrimination. Stiffer skin produces weaker deformation cues, forcing individuals to rely more on force and to use compensatory exploratory strategies.
- **Dynamic Cue Reweighting:** The brain does not rely on a fixed set of cues. Individuals dynamically reweight available information based on their skin properties and the task. For example, those with soft skin heavily use

(continued)

deformation cues (contact area, curvature), while those with stiff skin rely more on force-related signals.

Ultimately, an individual's tactile reality is a product of their unique biomechanical interface. Recognizing these sources of variation is crucial for developing adaptive haptic technologies and inclusive designs that are perceptually faithful across the entire user spectrum.

3.6 Skin Mechanics as a Lens That Shapes Perceptual Outcomes

Among these measured factors, skin stiffness emerged as the dominant determinant of compliance discrimination performance [24]. While finger size and fingerprint ridge density introduce measurable variability, statistical modeling shows that stiffness accounts for the largest portion of the variance in perceptual thresholds across the observed population. Individuals with lower skin stiffness consistently demonstrated higher perceptual sensitivity and greater accuracy in differentiating between compliance pairs. This finding highlights the central role of skin mechanics in shaping tactile perception and underscores its importance as a design variable in haptic systems.

3.6.1 Direct Modulation of Skin Mechanics

To further investigate how changes in skin stiffness influence tactile perception, experiments have been conducted to directly manipulate skin stiffness and observe the corresponding effects on perceptual acuity [18, 24]. By altering the mechanical properties of the skin, we can assess how different levels of stiffness affect mechanoreceptor responses, skin deformation patterns, and ultimately, the ability to perceive tactile stimuli (e.g., compliance). One method to change skin stiffness is through venous occlusion that increases finger pulp compliance. The findings reveal that increasing the compliance of skin does not alter the firing properties of SA-I afferents, which are primarily responsible for encoding fine spatial details. In contrast, RA-I afferents, which respond to dynamic tactile stimuli and low-frequency vibrations, exhibit reduced sensitivity under varied conditions of skin compliance. Furthermore, SA-II afferents, which are associated with skin stretch and shear force, also showed significantly reduced firing rate during venous occlusion, suggesting that increased skin compliance dampens the ability of these mechanoreceptors to encode large-scale deformations [18].

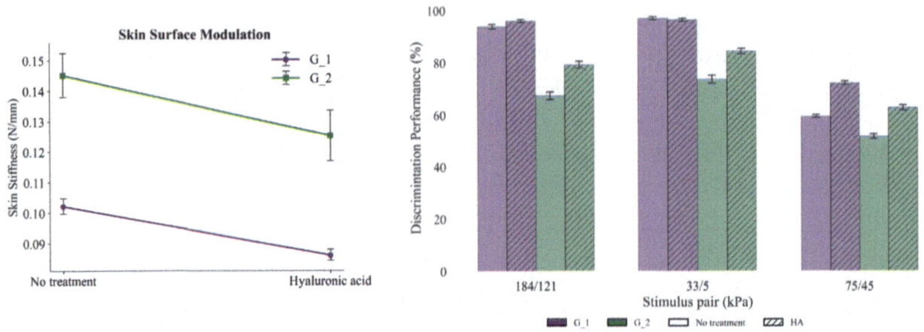

Fig. 3.4 Application of hyaluronic acid (HA) significantly reduces skin stiffness across participants. Correspondingly, both groups exhibit improved compliance discrimination performance after treatment. These results demonstrate that modulating skin mechanics directly enhances tactile perceptual sensitivity. G_1 and G_2 correspond to Group 1 and Group 2, respectively

Beyond affecting mechanoreceptor responses, altering skin stiffness also influences the overall deformation of the skin surface, which in turn modulates tactile perception. Complementarily, another method to alter skin stiffness is applying hyaluronic acid to soften the skin. Studies show that this softening effect leads to immediate and systematic changes in the skin's deformation patterns, generating pronounced biomechanical cues during contact, which are reflected in improved perceptual performance in discriminating between stimuli of varying compliance, especially for those participants with stiffer fingers, see Fig. 3.4. The findings relating skin stiffness to tactile acuity align with neurophysiological studies as well as computational models. Modulation of skin stiffness significantly influences the firing rates of cutaneous mechanoreceptors and finite element models show that this intervention modulates concentrations of stresses and strains near interior locations of mechanosensitive end organs [11, 17].

3.7 Conclusion

The variability of human tactile perception is not a boundary condition but rather defining of the design space for haptic systems. Differences in skin mechanics, receptor density, and perceptual thresholds mean that there is no single "typical" user for whom a haptic experience can be optimized. This diversity demands that hardware, software, and application subsystems be conceived not as fixed modules, but as adaptable components of a unified perceptual architecture. Each must anticipate and respond to variability rather than average it out.

Recognizing variability as intrinsic to touch reframes the goals of subsystem design. Hardware must provide the dynamic range and resolution needed to capture the full spectrum of human tactile behavior. Software must transform heterogeneous, user-dependent signals into perceptually consistent experiences, calibrating its models to individual differences in sensitivity and biomechanics.

When combined, these capabilities establish the foundation for closed-loop haptic systems that, like their biological counterparts, regulate perception through continuous adaptation.

In this sense, the second foundational principle extends beyond psychophysics into design practice: variability is not noise to be eliminated, but information to be leveraged. It is through this adaptability—across users, contexts, and materials—that integrated haptic systems can achieve perceptual robustness. The chapters that follow take up this challenge of translating variability into design: first in the hardware that captures touch, then in the software that interprets it.

References

1. Abdouni, A., Djaghloul, M., Thieulin, C., Vargiolu, R., Pailler-Mattei, C., Zahouani, H.: Biophysical properties of the human finger for touch comprehension: influences of ageing and gender. R. Soc. Open Sci. **4**(8), 170321 (2017)
2. Abdouni, A., Moreau, G., Vargiolu, R., Zahouani, H.: Static and active tactile perception and touch anisotropy: aging and gender effect. Sci. Rep. **8**(1), 14240 (2018)
3. Abdouni, A., Vargiolu, R., Zahouani, H.: Impact of finger biophysical properties on touch gestures and tactile perception: Aging and gender effects. Sci. Rep. **8**(1), 12605 (2018)
4. Amaied, E., Vargiolu, R., Bergheau, J.M., Zahouani, H.: Aging effect on tactile perception: Experimental and modelling studies. Wear **332**, 715–724 (2015)
5. Bouche, P., Cattelin, F., Saint-Jean, O., Leger, J., Queslati, S., Guez, D., Moulonguet, A., Brault, Y., Aquino, J., Simunek, P.: Clinical and electrophysiological study of the peripheral nervous system in the elderly. J. Neurol. **240**, 263–268 (1993)
6. Corniani, G., Saal, H.P.: Tactile innervation densities across the whole body. J. Neurophysiol. **124**(4), 1229–1240 (2020)
7. Dahiya, R.S., Gori, M.: Probing with and into fingerprints. J. Neurophysiol. **104**(1), 1–3 (2010)
8. Dahiya, R.S., Metta, G., Valle, M., Sandini, G.: Tactile sensing—from humans to humanoids. IEEE Trans. Robot. **26**(1), 1–20 (2009)
9. Deflorio, D., Di Luca, M., Wing, A.M.: Skin properties and afferent density in the deterioration of tactile spatial acuity with age. J. Physiol. **601**(3), 517–533 (2023)
10. Dillon, Y.K., Haynes, J., Henneberg, M.: The relationship of the number of meissner's corpuscles to dermatoglyphic characters and finger size. J. Anatomy **199**(5), 577–584 (2001)
11. Gerling, G.J.: Sa-i mechanoreceptor position in fingertip skin may impact sensitivity to edge stimuli. Appl. Bionics Biomech. **7**(1), 19–29 (2010)
12. Gerling, G.J., Thomas, G.W.: The effect of fingertip microstructures on tactile edge perception. In: First Joint Eurohaptics Conference and Symposium on Haptic Interfaces for Virtual Environment and Teleoperator Systems. World Haptics Conference, pp. 63–72. IEEE (2005)
13. Gerling, G.J., Thomas, G.W.: Fingerprint lines may not directly affect sa-i mechanoreceptor response. Somatosensory Motor Res. **25**(1), 61–76 (2008)
14. Gescheider, G.A., Bolanowski, S.J., Pope, J.V., Verrillo, R.T.: A four-channel analysis of the tactile sensitivity of the fingertip: frequency selectivity, spatial summation, and temporal summation. Somatosensory Motor Res. **19**(2), 114–124 (2002)
15. Gibson, G.O., Craig, J.C.: The effect of force and conformance on tactile intensive and spatial sensitivity. Exp. Brain Res. **170**, 172–181 (2006)
16. Gutiérrez-Redomero, E., Alonso, M., Dipierri, J.: Sex differences in fingerprint ridge density in the mataco-mataguayo population. Homo **62**(6), 487–499 (2011)

17. Hamasaki, T., Yamaguchi, T., Iwamoto, M.: Estimating the influence of age-related changes in skin stiffness on tactile perception for static stimulations. J. Biomech. Sci. Eng. **13**(1), 17–00575 (2018)
18. Hudson, K.M., Condon, M., Ackerley, R., McGlone, F., Olausson, H., Macefield, V.G., Birznieks, I.: Effects of changing skin mechanics on the differential sensitivity to surface compliance by tactile afferents in the human finger pad. J. Neurophysiol. **114**(4), 2249–2257 (2015)
19. Jarocka, E., Pruszynski, J.A., Johansson, R.S.: Human touch receptors are sensitive to spatial details on the scale of single fingerprint ridges. J. Neurosci. **41**(16), 3622–3634 (2021)
20. Kalisch, T., Ragert, P., Schwenkreis, P., Dinse, H.R., Tegenthoff, M.: Impaired tactile acuity in old age is accompanied by enlarged hand representations in somatosensory cortex. Cerebral Cortex **19**(7), 1530–1538 (2009)
21. LaMotte, R.H., Srinivasan, M.A.: Surface microgeometry: Tactile perception and neural encoding. In: Information Processing in the Somatosensory System: Proceedings of an International Symposium at the Wenner-Gren Center, Stockholm, 3–5 July, 1989, pp. 49–58. Springer (1991)
22. Lévêque, J.L., Dresler, J., Ribot-Ciscar, E., Roll, J.P., Poelman, C.: Changes in tactile spatial discrimination and cutaneous coding properties by skin hydration in the elderly. J. Investigative Dermatol. **115**(3), 454–458 (2000)
23. Li, B., Gerling, G.J.: Individual differences impacting skin deformation and tactile discrimination with compliant elastic surfaces. In: 2021 IEEE World Haptics Conference (WHC), pp. 721–726. IEEE (2021)
24. Li, B., Gerling, G.J.: An individual's skin stiffness predicts their tactile discrimination of compliance. J. Physiol. **601**(24), 5777–5794 (2023)
25. Li, B., Hauser, S., Gerling, G.J.: Identifying 3-d spatiotemporal skin deformation cues evoked in interacting with compliant elastic surfaces. In: 2020 IEEE Haptics Symposium (HAPTICS), pp. 35–40. IEEE (2020)
26. Li, B., Hauser, S.C., Gerling, G.J.: Faster indentation influences skin deformation to reduce tactile discriminability of compliant objects. IEEE Trans. Hap. **16**(2), 215–227 (2023)
27. Loesch, D., Martin, N.: Finger ridge patterns and tactile sensitivity. Ann. Human Biol. **11**(2), 113–124 (1984)
28. Loesch, D., Martin, N.: Relationships between minute characteristics of finger ridges and pattern size and shape. Ann. Human Biol. **11**(2), 125–132 (1984)
29. McIntyre, S., Nagi, S.S., McGlone, F., Olausson, H.: The effects of ageing on tactile function in humans. Neuroscience **464**, 53–58 (2021)
30. Mountcastle, V.B.: The sensory hand: neural mechanisms of somatic sensation. Harvard University Press, Cambridge (2005)
31. Palve, S.S., Palve, S.B.: Impact of aging on nerve conduction velocities and late responses in healthy individuals. J. Neurosci. Rural Practice **9**(1), 112 (2018)
32. Peters, R.M., Goldreich, D.: Tactile spatial acuity in childhood: effects of age and fingertip size. PloS One **8**(12), e84650 (2013)
33. Peters, R.M., Hackeman, E., Goldreich, D.: Diminutive digits discern delicate details: fingertip size and the sex difference in tactile spatial acuity. J. Neurosci. **29**(50), 15756–15761 (2009)
34. Pham, T.Q., Hoshi, T., Tanaka, Y., Sano, A.: Effect of 3d microstructure of dermal papillae on sed concentration at a mechanoreceptor location. PloS One **12**(12), e0189293 (2017)
35. Phillips, J., Johansson, R.S., Johnson, K.O.: Responses of human mechanoreceptive afferents to embossed dot arrays scanned across fingerpad skin. J. Neurosci. **12**(3), 827–839 (1992)
36. Prevost, A., Scheibert, J., Debrégeas, G.: Effect of fingerprints orientation on skin vibrations during tactile exploration of textured surfaces. Communicative Integrative Biol. **2**(5), 422–424 (2009)
37. Scheibert, J., Leurent, S., Prevost, A., Debrégeas, G.: The role of fingerprints in the coding of tactile information probed with a biomimetic sensor. Science **323**(5920), 1503–1506 (2009)
38. Srinivasan, M.A., LaMotte, R.H.: Tactual discrimination of softness. J. Neurophysiol. **73**(1), 88–101 (1995)

39. Vega-Bermudez, F., Johnson, K.O.: Fingertip skin conformance accounts, in part, for differences in tactile spatial acuity in young subjects, but not for the decline in spatial acuity with aging. Percept. Psychophys. **66**(1), 60–67 (2004)
40. Verrillo, R., Bolanowski, S., Francis, C., McGlone, F.: Effects of hydration on tactile sensation. Somatosensory Motor Res. **15**(2), 93–108 (1998)
41. Xu, C., Wang, Y., Gerling, G.J.: Individual performance in compliance discrimination is constrained by skin mechanics but improved under active control. In: 2021 IEEE World Haptics Conference (WHC), pp. 445–450. IEEE (2021)
42. Yang, Y., Wang, L., Yan, F., Xiang, X., Tang, Y., Zhang, L., Liu, J., Qiu, L.: Determination of normal skin elasticity by using real-time shear wave elastography. J. Ultrasound Med. **37**(11), 2507–2516 (2018)

Chapter 4
From Scientific Foundations to Hardware Design Goals

Abstract With the scientific foundations of touch established, this chapter begins the transition from understanding to building—from the natural mechanics of skin and perception to the engineered systems that seek to replicate them. It considers how the physical laws that govern deformation and sensation can be translated into the design principles of hardware, and how the quality of what is sensed ultimately shapes what can be known. Rather than treating sensors as instruments of measurement alone, the chapter frames them as the physical layer of an inference process: as devices that must not only detect mechanical events, but also encode them in ways that preserve perceptual meaning. By tracing the evolution of haptic hardware and examining the shift from measurement-centric to inference-centric design, the chapter positions hardware as the source of the informational substrate, data, upon which software and application subsystems will build.

4.1 Hardware as the Foundation of Integrated Haptic System Design

As haptic science deepens our understanding of how tactile information is physically generated and perceptually interpreted, it offers a powerful foundation for engineering design. The two scientific foundations established in the first half of this book—that skin mechanics and psychophysics define the operating envelope of touch, and that individual differences shape what is perceived—point toward a more systematic, perception-driven approach to designing haptic technologies.

The first foundation established the biomechanical aspect of touch. It revealed how skin mechanics and psychophysics jointly define the physical pathways through which tactile information is generated, transmitted, and encoded. It defines the operating range of the human tactile perception, describing how mechanical events occurred at the skin give rise to perceptually relevant cues, and how those cues are transformed into perceptual representations. From a design perspective, this foundation defines the mechanical boundary conditions for hardware: the resolutions, dynamic ranges, and stimulus dimensions that must be captured to remain aligned with the physics of touch and the thresholds of human perception.

B. Li, T. Cody, *Design Principles for Integrated Haptic Systems*, Springer Series on Touch and Haptic Systems, https://doi.org/10.1007/978-3-032-16986-0_4

The second foundation established the informational aspect of touch. It extended the biomechanical baseline by addressing the variability of tactile perception—how anatomical, skin mechanical, and cognitive differences among individuals give rise to distinct sensory representations and perceptual interpretations. It reframed touch not as a uniform process, but as a personalized, adaptive, and context-dependent mode of information processing. From an engineering viewpoint, this foundation specifies the informational boundary conditions: how haptic systems must represent, adapt, and personalize tactile information to account for individual diversity.

These two foundations define the physical substrate and the informational logic of haptic perception. The first tells us how tactile information originates through biomechanical interaction, and the second emphasizes how that information is organized and interpreted through perceptual and cognitive variability. They transform haptic design from a problem of measurement to a problem of inference: not merely detecting what happens at the surface, but inferencing what it means in perceptual and functional terms.

From the integrated haptic system design process perspective, in an inference-centric haptic system, hardware, software, and application subsystems are stages of an inference pipeline—a process that transforms raw mechanical cues into perceptual and contextual understanding. Within this pipeline, the hardware subsystem forms the foundation: it determines the fidelity, quality, dimensionality, and interpretability of the sensory data upon which all higher-level inference depends. Hardware thus defines the system's capacity to know and reason about the tactile world. If the signals it produces are incomplete, distorted, or poorly structured with respect to perceptual variables, no downstream computation can fully reconstruct the lost meaning.

This chapter marks a transition of the book: from understanding human touch, to exploring how that understanding can be harnessed to design systems that behave in more human-aligned, perceptually meaningful ways. Here, we begin to move from science to engineering and from theory to implementation. The remainder of this chapter starts with a historical overview of haptic hardware development to reveal how past progress has been limited by a measurement-centric paradigm, and then articulates the emerging design goals for inference-centric, science-grounded hardware subsystems. Using compliance perception as a case study, we demonstrate how the science of compliance—rooted in skin biomechanics and psychophysics—can be translated into concrete hardware design goals that support perceptual inference. This shift from measurement to inference re-centers hardware around the principles of perceptual relevance and inference utility, moving beyond component-level optimization toward systems-oriented, perception-aligned hardware design that will serve as the foundation for the following chapters on software and application subsystems.

4.2 Technological Evolution of Haptic Hardware

The development of haptic hardware has evolved with increasing sophistication in both form and function [24, 38, 49], see Fig. 4.1. Early haptic hardware was built around single-point sensing and actuation mechanisms, using technologies such as load cells, strain gauges, and vibration motors to detect and simulate tactile input [7, 10, 16]. These devices, often embedded in VR controllers or touchscreen interfaces, provide localized feedback but lack spatial resolution and biomechanical realism [17, 58]. For example, haptic gloves and VR controllers have been developed with force sensors and vibration motors placed on fingertips and the palm to detect contact force and provide basic vibrational cues in virtual environments, offering low-cost, simple implementations but are limited to a single axis of interaction [37, 39, 44, 50]. Touchscreen interfaces and virtual buttons, implemented using either piezo actuators or vibration motors, also adopted single-point feedback to enhance user experience and convey directional cues for nonvisual navigation, though the signals are abstract, confined to vertical contact cues, and lacked perceptual richness [3, 5, 20, 20, 35]. While these devices using single-point sensing or actuation mechanism can effectively signal discrete events, they fail to capture the rich, distributed tactile cues that are essential for naturalistic sense of touch.

These limitations lead to the emergence of haptic hardware that utilizes multi-point array or pixel-based designs, which significantly improve spatial resolution and functional richness [6, 19, 57]. These devices are composed of densely arranged sensing elements—often referred to as tactile pixels or taxels—that enable localized detection and response across a surface. Unlike single-point designs, multi-array architectures can detect not only the magnitude but also the direction of applied

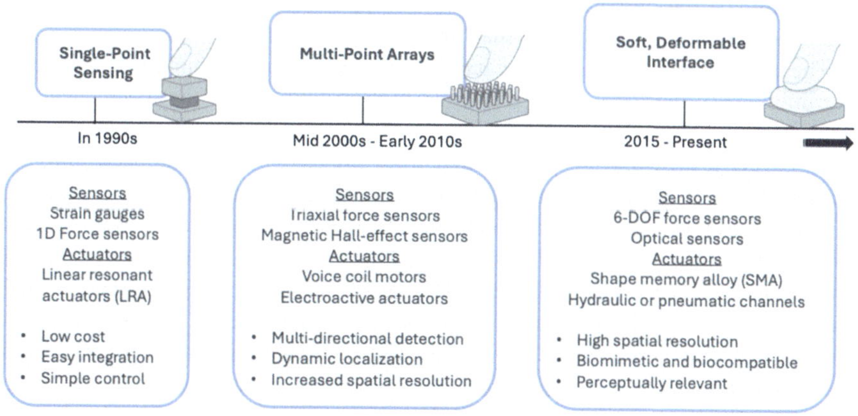

Fig. 4.1 Technological evolution of haptic hardware devices. This progression reflects an increasing alignment between device capabilities and the mechanical and perceptual properties of human skin

forces, supporting multi-axial sensing through the use of triaxial or 6-degree-of-freedom (6-DoF) force and torque sensors for haptic rendering [9, 36, 46]. Some devices are also capable of recognizing both static contact and dynamic events, allowing for more robust temporal interpretation of tactile interactions [1, 34, 51]. While these devices offer improved functionality for applications such as shape recognition, object manipulation, and distributed contact sensing, several key limitations remain. First, even with high-density sensing arrays, the spatial resolution of most devices remains significantly lower than that of human skin, which contains mechanoreceptor spacing on the order of 1-2 millimeters in the fingertip. Second, although these devices can estimate surface dynamics, they generally lack the capability to capture high-resolution, quantitative measurements of skin deformation, which are critical biomechanical cues underpinning tactile perception. Third, like most single-point sensing devices, multi-array designs often rely on rigid materials or incorporate embedded rigid components, limiting their ability to conform to complex geometries or integrate seamlessly with the human body or robotics. Fourth, due to the absence of perceptually meaningful quantitative metrics, performance evaluations typically depend on subjective user ratings, hindering reproducibility and scientific validation. Fifth, as the number of taxels increases, so do the demands on power consumption and signal processing, creating challenges for real-time performance and energy efficiency. Finally, multi-array designs are prone to cross-talk and noise interference between adjacent sensing units, which can compromise measurement accuracy and fidelity[42, 43]. These limitations underscore that, while multi-point contact mechanism represents a significant step forward in haptic technology, they still fall short in capturing high-resolution biomechanical cues. To more faithfully sense and replicate the nature of touch, the next generation of haptic hardware must be both mechanically analogous—soft, conformable, and elastic—and perceptually capable extracting high-dimensional, quantitative cues of 3D skin deformation patterns. These capabilities are essential for enabling closed-loop haptic feedback and supporting more intelligent, embodied interactions between humans and machines.

The current state-of-art in haptic hardware marks a turning point toward systems that utilize soft, deformable materials and high-resolution sensing mechanisms to better replicate the mechanical and functional characteristics of human skin. This shift reflects a growing recognition that truly effective haptic feedback must be grounded in both technological capability and also biomechanical and perceptual realism. Advanced sensing techniques have emerged as powerful tools in this domain. For example, optical-based sensors enable high spatial and temporal resolution, and can capture subtle deformations of soft surfaces through visual cues such as reflected light patterns or marker displacement [26, 41, 60]. With advancements in experimental paradigms and computational tools, cameras and vision-based systems are now capable of observing complex skin deformation patterns and extracting biomechanical features that are strongly correlated with human tactile perception [28, 30, 32]. This evolution in hardware aligns closely with the trajectory of haptic science, which has progressed from studying simple one-dimensional cues (e.g., pressure or force magnitude) to more sophisticated two-dimensional cues such as surface strain and displacement, and is now moving toward a more complete characterization of 3D tactile cues. These include penetration depth, surface curvature, dynamic contact area, and contact shape, all of which are important for conveying material properties (e.g., compliance), object geometry, and interaction dynamics in a perceptually meaningful way. However, very few existing haptic devices are designed to reliably capture or replicate these high-dimensional biomechanical patterns. This gap suggests that future haptic hardware should not only simulate force or vibration, but also encode and reproduce perceptually correlated skin deformation cues with high fidelity.

4.2.1 Transformation of Haptic Hardware

The implications of this transformation from single-point actuation to spatially distributed arrays, and now to soft, skin-like, high-resolution interfaces, extend beyond hardware alone. By integrating quantitative, high-dimensional, perceptually grounded deformation data into hardware design, these future haptic systems establish a critical foundation for advanced software architectures and integrated solutions that are adaptable, intelligent, and scalable. As sensors evolve to capture detailed 3D deformation patterns and real-time physiological interactions at the skin-device interface, they unlock transformative capabilities in tactile perception modeling, controller adaptation, and system-level efficiency.

Predictive Modeling with High-dimensional Data High-resolution, camera-based tactile sensors, such as vision-based elastic gel pads, can stream dense deformation profiles in real-time, producing in order of thousands vector readings per second [21, 61]. This spatiotemporal richness powers machine learning models that can infer critical contact-related properties, including object identity, slip onset, force distribution, and grasp stability: capabilities that are out of reach for traditional one-axis load cell or low-resolution taxel arrays.

Personalization Through Measured Skin Biomechanics By measuring how each user's skin deforms under contact, soft sensors make it possible to estimate individual-specific skin properties, such as stiffness and conformance [15, 29, 56]. These data enable controllers to learn personalized transfer functions that modulate actuation gains based on the user's unique biomechanical profile: e.g., applying higher force output for stiffer fingers to improve their perceptual performance. Additionally, these data can be used to monitor age-related or hydration-induced changes in skin layers over time, enabling systems to dynamically recalibrate based on users' current skin state beyond general, one-size-fits-all models. Such individual adaptation is critical for both usability and accessibility, particularly in medical, prosthetic, or aging-related applications.

Closed-loop Feedback for Real-time Adaption High-resolution skin deformation sensing also supports the implementation of closed-loop feedback systems, where real-time user response directly informs actuation control [45, 48]. Rather than relying on pre-programmed haptic patterns or open-loop feedback, the system adjusts continuously based on ongoing measurements of skin deformation cues. This enables more stable, context-sensitive interactions, such as modulating grip force in response to slip onset or adapting vibrotactile signals based on user fatigue or motion trajectory.

System-level Simplification and Scalability High-dimensional tactile interfaces simplify upstream processing and enable broader scalability. For example, when each fingertip module or sensor patch outputs pre-calibrated, high-fidelity 3D tactile information, system architecture becomes lighter and more modular. Instead of depending on centralized fusion of raw, noisy signals, higher-level applications can operate on compressed, interpretable tactile representations. This can be largely benefited from extracting perceptually related skin deformation cues by the hardware devices[8, 30, 32]. Moreover, the modularity and decentralized processing mechanism reduce computational overhead and support easier extension from single-finger systems to whole-hand or whole-body tactile coverage, which is essential for scalable wearable interfaces or robotic skin.

4.2.2 Scientific and Technological Limitations

Even as hardware becomes more capable, current haptic systems still struggle to capture and reproduce the high-dimensional skin deformation cues that underlie human tactile perception. Touch is not solely about force and vibration—it involves rich spatiotemporal deformation patterns shaped by the biomechanical properties of the skin and the interaction dynamics with object surfaces. These cues are essential for perceiving compliance, texture, and shape, as discussed in Chap. 2. Without the ability to sense and replicate these patterns, haptic devices cannot deliver perceptually accurate or functionally useful feedback.

Moreover, many haptic systems continue to rely heavily on psychophysical evaluations that depend on subjective user feedback. This reliance on subjective assessments obscures the understanding of how specific physical interactions map onto perceptual outcomes, and also hinders systematic comparisons across devices, users, and experimental conditions. The absence of standardized, quantitative metrics grounded in biomechanics or neural encoding makes it difficult to assess the fidelity or effectiveness of haptic interfaces in a consistent and objective manner. Without measurable outcomes such as skin deformation profiles, mechanoreceptor activation models, or psychophysical thresholds tied to biomechanical inputs, haptic hardware development lacks a robust framework for benchmarking and innovation.

This gap—the lack of quantitative sensing of skin biomechanics and measurements of user's psychophysics—also constrains the advancement of closed-loop haptic systems. Without accurate, real-time measurements of skin deformation and a model linking those biomechanical inputs to perceptual performance, systems remain open-loop, delivering static or preprogrammed feedback that fails to adapt to dynamic, user-specific responses. While force and displacement are often measured, they are insufficient to fully characterize tactile interactions, especially in contact involving soft, textured, or dynamic stimuli, as described in Chap. 2.

These limitations persist not simply because of technological constraints, but more importantly due to a disconnect between engineering practice and scientific understanding of touch. Bridging this gap requires science-based approach: one that translates quantitative measurements from skin biomechanics, tactile psychophysics, and neural encoding into engineering design specifications. Such a framework will address these limitations in haptic hardware and also establish a pathway for building haptic systems that are quantitatively validated, perceptually aligned, and capable of personalized, adaptive interaction.

4.3 Inference-Centric Hardware Design for Compliance Perception

Compliance perception, the ability to perceive softness or hardness of materials, provides bridge between skin mechanics and psychophysics, showing how mechanical interaction at the skin surface gives rise to perceptual meaning. When a fingertip interacts with an object, the skin undergoes complex deformations: indentation depth increases, lateral stretch develops, surface curvature changes, and contact area expands or contracts. These deformation patterns constitute a dynamic stream of information. These physical interactions define what the hardware can measure, how the software can interpret it, and ultimately how the system as a whole can infer material properties.

In inference-centric haptic systems, the hardware subsystem functions not merely to measure force or displacement, but also as an informational interface. Its functional role is to generate high-quality, perceptually interpretable data that preserve the structure of the underlying biomechanical event. The fidelity and dimensionality of this data determine the system's inferential capacity—what perceptual variables it can reconstruct, how finely it can discriminate between materials, and how reliably it can predict user experience.

The remainder of Sect. 4.3 will explore how to translate scientific principles into concrete performance targets and design strategies for hardware subsystems that support human-level compliance perception in integrated haptic systems.

4.3.1 Prioritizing Deformation-Based Cues

Traditional hardware often aims to replicate compliance by matching force-displacement curves or simulating terminal force. But research has consistently shown that these cues are insufficient on their own. Instead, humans primarily rely on spatiotemporal skin deformation cues—such as contact area evolution, surface curvature, penetration depth, and lateral strain patterns—to infer compliance [2, 4, 8, 30, 40]. Empirical studies show that cues such as contact area expansion, penetration depth, and curvature change dominate perceptual judgments, especially in soft regimes. In contrast, cues like force rate and contact shape are more informative to compliance discrimination in stiffer materials [13, 27, 59].

These cues are spatially distributed and temporally evolving, and they activate mechanoreceptor populations in distinct, information-rich ways. From an design standpoint, these cues represent high-informational channels—variables that carry the richest signal-to-noise ratio for compliance inference. Hardware subsystems should therefore be optimized to capture or reproduce these deformation cues explicitly with sufficient spatial and temporal resolution [11, 31, 33]. Rather than focusing on absolute accuracy of force rendering, the design objective shifts toward the informational quality of deformation cues that directly support human-like inference.

4.3.2 Psychophysical Thresholds and Perceptual Design Targets

Psychophysical studies provide quantitative boundaries for perceptual inference, defining thresholds that hardware systems should meet or exceed to achieve perceptual fidelity. For example, in compliance perception, Weber fractions provide quantifiable targets: users can discriminate approximately 13–15% changes in compliance when skin deformation cues are present, but the threshold increases to 50% or more when relying on force-only input [47, 55]. Similarly, psychophysical studies show that subjects lacking cutaneous input (via finger anesthesia) could not discriminate rubber samples differing 90% in compliance, whereas with only cutaneous cues, they could discriminate less than 75% differences in material compliance [22, 47]. In line with these findings, some tasks even show single-finger pressing outperforming two-finger pinching in compliance differentiation [12, 52, 53], illustrating how consistent, localized cutaneous information can surpass more complex force inputs. Taken together, these psychophysical thresholds reinforce the critical importance of surface deformation in achieving human-level compliance discrimination

These values can be recast as performance metrics for subsystem design: the sensing and actuation mechanisms must achieve a spatial deformation resolution and temporal update rate that preserves such perceptual distinguishability in compliance. A system that can capture or simulate skin deformation with fidelity approaching this threshold can enable perceptual inference that mirrors natural touch, while systems limited to gross force feedback will impose a higher perceptual workload and lower discrimination accuracy. Thus, perceptual thresholds serve not only as validation metrics but also as design constraints for sensing resolution, actuation fidelity, and control bandwidth.

4.3.3 Interaction Mode Shapes Sensing Priorities

Whether contact is direct (bare-finger) or indirect (mediated) defines the sensory pathways and thus the informational requirements for hardware. In bare-finger interaction, the perception of compliance derives primarily from cutaneous cues generated by skin deformation-how skin surface expands, curves, and stretches against an object. Accordingly, systems that aim to replicate softness or hardness must focus on surface-deformation cues such as contact area, penetration depth, or skin curvature. This is particularly relevant when simulating soft objects, which produce large, gentle, and slower-developing deformation patterns, versus stiff objects, which constrain skin deformation to a more 2D representation with a rapid change in contact shape and ramping force more abruptly. Hence, a haptic device for direct touch should be equipped to capture and replicate the appropriate spatiotemporal pattern of skin deformation for effective simulation of compliance sensation.

In contrast, when a tool is used to mediate touch, much of the tactile information shifts from direct cutaneous signals on the skin surface to force and vibration cues transmitted along the tool's shaft. Under these conditions, the device's capacity to present time-varying signals becomes critical: a rapid raise in force (or high-frequency vibration bursts) can signal hardness or stiffness, while a more gradual buildup suggests softness. Because users rely heavily on these transients, rather than on end-point force or vibration magnitude, to discern different levels of compliance, static or steady-state measurements alone cannot convey the dynamic nature of compliance perception. Devices must therefore ensure a sufficient bandwidth to render the quick variations in force and vibrotactile feedback that differentiate soft from stiff materials. Whether the tool is tapped, pressed, or lightly probed, its force output (and any vibratory "ringing") should be updated in nearly real-time to reflect the underlying properties of the virtual or remote object, capturing the instantaneous changes that human perception requires for a realistic sense of compliance.

Designing hardware that performs effectively across both direct and mediated interactions requires more than mechanical precision; it requires context-sensitive encoding—the ability to identify and represent the physical variables that carry perceptual significance in each mode of contact. In one context, those variables may be patterns of skin strain; in another, they may be temporal profiles of force and vibration. Context-sensitive encoding allows the system to adapt its sensing and actuation strategy to the informational structure of the interaction, emphasizing what matters most to perception rather than reproducing forces indiscriminately. This shift marks a transition from fixed mechanical replication to inference-oriented design, where hardware operates not merely as a transmitter of signals, but as a selective, perceptually aligned interpreter of tactile events.

4.3.4 *Mechanical and Temporal Ranges for Realistic Interaction*

To enable perceptual realism and accurate inference, hardware systems must operate within realistic mechanical and temporal ranges. Empirical studies establish the operational range within which compliant interactions are both biomechanically valid and perceptually informative. A critical dimension of haptic device design is the range and resolution of force input and output. Empirical studies show that people typically apply near 2 to 3 Newtons (N) of force in everyday manipulations, but may reach 5 N or more to distinguish stiffer objects. Force-rate cue, with 0.1 N interval, shows statistically significant correlation with compliance discrimination [13], indicating that providing a 0 to 5 N force range (or higher) with at least 0.1 N resolution can replicate these everyday force profiles and capture subtle force-related cues that used to differentiate compliance levels. In addition, the device's ability to modulate force ramps at a sufficient control bandwidth is vital, because quick changes in force (i.e., force rate) strongly influence perceptual judgments of compliance [22].

Beyond force, contact parameters such as indentation displacement and contact angle also shape the realism of a device. Natural finger-object interactions often involve indentations of 1–5 millimeters, though even <2 millimeters can be sufficient if velocity and force are controlled [30]. Designing a system to allow this range of normal compression enables more faithful simulation or reproduction of how the fingertip actually deforms, thereby drive realistic sense of compliance. A 30° contact angle relative to the contact surface has been adopted widely in haptic science as it activates most of the mechanoreceptors concentrated at the fingertip. Moreover, directions of contact (normal vs. tangential) should also be considered in incorporating into the design: while normal compression emphasizes depth and curvature cues, tangential forces activate shear and skin-stretch pathways essential for certain textures and softness distinctions [4].

The velocity of contact is similarly crucial as it directly influences neuron firing patterns and discharge rates. In active exploration, users can spontaneously vary contact velocity between approximately 1–20 mm/s, adapting their speed as they gather feedback [54]. In passive touch, the velocity is controlled (often between 1–15 mm/s), yet participants still rely on similar skin deformation cues even without any kinesthetic input. From the steady-state vibratory testing, the psychophysical sensory channels respond to vibration with sensitivity up to 300 Hertz(Hz), however, practically a device update rate of 10–200 Hz for force or displacement rendering can help represent the rapid transients that differentiates compliance variations [23, 30]. Equally important is the time duration of contact. Studies show that >0.4 s is required to perceptually detect differences in compliance, even though biomechanical differences start to appear before that [32].

For subsystem design, these parameters define the operating envelope of compliance capability in the haptic systems. They set integration constraints: hardware sensing, actuation, and control loops must operate within these limits to maintain perceptual coherence and interoperability with higher-level software modules. By allowing adequate sensing or actuation parameters such as force range, displacement, and controlling contact conditions such as angles and velocities over sufficient contact time duration, designers can ensure the reliable simulation of critical spatiotemporal signals that guide naturalistic tactile discrimination experiences. Meeting these parameters ensures that data streams entering the system remain aligned with human perceptual timing and mechanical sensitivity.

4.3.5 Neural Coding Inspires Multi-Channel Design

Biological tactile systems infer compliance by integrating inputs from multiple mechanoreceptor populations. Multiple skin deformation cues, such as contact area, curvature, penetration depth, and force, activate diverse populations of mechanoreceptors, each tuned to specific aspects of physical stimulation. Slowly adapting type afferents, for instance, are sensitive to sustained pressure and spatial variations in deformation, effectively capturing the magnitude of contact area or how deeply the skin penetrates into the material. In contrast, rapidly adapting afferents respond robustly to dynamic or vibratory stimuli, such as abrupt changes in force. By parsing information in parallel, these receptor types function as distinct yet complementary channels of input: each cue, be it spatial or temporal, sustained or transient, provides an independent source of data on perception of compliance.

Despite this independence, these cues also reinforce one another to enrich compliance perception. An SA-I afferent, for example, may escalate its firing rate when local skin curvature intensifies, indicating increased pressure on a particular skin region, while an RA afferent in the same region may spike in response to a rapid expansion of contact area. Meanwhile, SA-II or deep mechanoreceptors (e.g., Pacinian corpuscles) can detect broader strain patterns and high-frequency vibrations, further refining the underlying percept. Through this division of labor, the sensory system distributes computational demands among specialized pathways; SA-I units focus on spatial details and local gradient changes, RA units on transients and rate signals, and SA-II afferents on global strain variation. This allows the central nervous system to integrate multiple streams of relevant data without overloading any single channel.

The result is a perceptual synergy: each cue has the capacity to independently encode facets of compliance, yet together they form the nuanced, context-sensitive percept of softness or hardness we rely on in everyday touch. As contact condition vary, such as changes in exploration strategy, velocity, or the material's modulus, the brain can re-weight these cues dynamically, adjusting the relative emphasis on static versus transient signals. In this way, the independent roles of each cue remain a cornerstone of tactile coding, but it is ultimately their complementary interplay that furnishes the richly detailed perception compliance.

Consequently, to build effective haptic devices that replicate such intricate perception of compliance, one key design principle is to offer multiple channels of feedback. Instead of a single force-displacement loop, systems can simultaneously provide information such as localized penetration depth, dynamic contact area change rate, or transient vibrations. Such parallel processing mirrors the division of labor in the nervous system, allowing each cue (spatial and temporal) to stand alone while also reinforcing the realism of the overall tactile experience. By splitting these cues among different actuation mechanisms or control loops, the device can avoid overloading any single signal path.

Another important implication for design arises from how the haptic system re-weights these cues depending on specific physical interactions. To reproduce different levels of compliance, softer objects may require large displacements and expansive contact area simulations, whereas stiffer objects can rely on a sharper change on force and contact shape. Designers can incorporate hardware and software modules that dynamically adjust these channels accordingly to real-time user input and feedback, thus aligning with the brain's strategy of selectively emphasizing certain cues on context basis.

A further consideration is the benefit of using fine-grained or distributed actuation mechanism. Because of the high spatial distribution of mechanoreceptors that are integral to cutaneous interpretation of compliance, a single-point actuator or uniform surface modulation may fail to capture the details the nervous system expects. Instead, allowing multiple, independently controlled actuation components can create defined differential shear or indentation across the skin contact region. This local variance better approximates the naturally occurring patterns of skin deformation that the corresponding mechanoreceptors can encode. Such distributed design increases informational redundancy and robustness, enabling the system to infer compliance even under variable contact conditions.

4.3.6 Compliance as a Template for Hardware Design

Within the integrated haptic system design framework, compliance perception becomes a case study in how science informs subsystem design. The biomechanical foundations of skin deformation define what must be physically represented, while the informational foundations of perception define how those representations must be structured for inference. The role of the hardware subsystem is thus to produce data that are mechanically valid, perceptually relevant, and computationally useful. Aligning hardware performance targets with psychophysical metrics, and ensuring that sensing and actuation operate within perceptually meaningful ranges, creates a direct bridge between human tactile capability and machine implementation. This translation from perceptual science to hardware design establishes the framework for the following chapters, which extend the same inference-centric logic to the software and application subsystems of integrated haptic systems.

4.4 Multifunctional and Multimodal Hardware Design for Inference

Building on the previous foundations, the next step is to define the design considerations for future haptic hardware: devices that not only simulate touch, but serve as intelligent, information-generating subsystems within an integrated haptic system. Human touch is both multimodal and multifunctional: it simultaneously detects mechanical pressure, shear, vibration, temperature, and even nociceptive or affective qualities such as pain or pleasantness, and it interprets these through multiple functional dimensions—direction, rate, friction, and dynamic change. To replicate such richness, haptic hardware must similarly sense and render across multiple dimensions, producing structured, high-quality data that support perceptual inference at the system-level.

4.4.1 Multi-Function Sensing

In the human tactile system, no single cue fully describes an interaction—pressure, shear, and slip interact dynamically to define contact. Future haptic hardware must therefore go beyond normal force measurement to transduce multiple mechanical dimensions: normal and tangential forces, shear direction, slip onset, contact rate, and vibration spectrum. Each of these parameters contributes a distinct informational channel for inference.

For example, differentiating compliance requires both force-displacement tracking and contact area change rate monitoring; detecting incipient slip demands sensitivity to both shear force direction and rapid tangential accelerations. Sensors and actuators, likewise, must render both sustained loads and transient cues, coordinating across multiple degrees of freedom. Such capabilities are critical for replicating the diverse tactile signals encountered during real-world interactions. To support these functions, hardware subsystems must offer high temporal resolution, multi-axis sensing, and spatially distributed architectures.

From an inference-centric perspective, the goal of multi-function sensing is to expand the feature space of tactile data so that software can infer higher-order percepts directly from physical events. By providing this multidimensional input, the hardware subsystem enables algorithms to reconstruct the context, material, and intent of touch.

4.4.2 Multi-Modal Rendering

Touch is inherently multimodal, integrating mechanical, thermal, and affective cues into a unified perceptual experience [14, 18, 25]. The human skin encodes not only pressure and vibration, but also temperature, humidity, and nociceptive signals that modulate how materials feel—soft or rigid, warm or cold, pleasant or irritating.

Incorporating these sensory modalities into hardware broadens the system's informational basis for inference. Thermal sensors and actuators, for instance, allow the system to capture and render thermal conductivity and heat flow, key to distinguishing metal from fabric. Humidity or wetness sensors add cues for slipperiness and stickiness, while nociceptive or pain-like signals extend feedback into the affective and protective domains.

Such multimodal integration provides the software layer with a richer data manifold for interpreting affective states and environmental qualities, improving realism and functional accuracy in applications like teleoperation, prosthetics, and affective communication, where tactile perception must reflect the full sensory spectrum of human skin.

4.4.3 Structuring the Interface for Skin-Like Deformation

To faithfully replicate the interface between human skin and real-world objects, haptic devices must exhibit mechanical properties that are close to those of biological tissue. This includes adopting materials and structural designs that approximate the compliance, curvature adaptability, and the ability to form a stable, wide-area contact under dynamic loading conditions of human skin. Soft, elastic materials with tunable stiffness and high conformity ensure that the device interface deforms correspondingly with the user's finger pad, preserving signal integrity for both actuation and sensing. Such biomimetic, skin-like mechanics improve perceptual realism by producing deformation patterns that align with those generated in actual human touch, and it also helps stabilize the contact interface, reduces motion artifacts, and enhances spatial resolution in tactile signal acquisition—making it optimal for driving predictive psychophysical models. As a result, these psychophysical outputs are interpretable to both human perception and computational models, minimizing mechanical mismatches and reducing noise.

4.4.4 Built-In Calibration for Contextual Variability

Because tactile perception varies widely across individuals—due to differences in skin stiffness, ridge density, and neural sensitivity—future hardware systems should incorporate built-in calibration mechanisms that personalize sensing and feedback. These may include one-time self-calibration routines or continuous adaptive algorithms that adjust gains, thresholds, or waveform parameters in real-time.

In an inference-centric haptic systems, calibration ensures that the hardware subsystem maintains consistent informational quality across users and contexts. By accounting for individual differences, the system preserves the comparability and interpretability of its tactile data, allowing higher-level modules to operate within a stable perceptual reference frame. This adaptivity transforms hardware from a static measurement device into a context-aware perceptual interface.

4.5 From Biophysical Input to Psychophysical Meaning

When future haptic hardware incorporates multi-functional transduction, multi-modal sensing, bio-inspired mechanics, and build-in calibration, it produces data that are not only mechanically rich but perceptually aligned—reflecting the tactile cues that human nervous system uses to interpret touch. These signals span multiple dimensions of tactile interaction, such as force magnitude in multiple directions, slip onset, and surface temperature. Crucially, this multi-dimensional input is what enables software and systems to bridge the gap between physical events and perceptual meaning.

Such high-fidelity, multi-channel data streams provide the raw material for advanced computational models that can predict user percepts based on real-time sensor input. For example, machine learning models can be trained to infer compliance or texture perception from skin deformation fields and vibration profiles, allowing the system to render haptic feedback that adapts seamlessly to the user's expectations and environment. These predictive models shift the paradigm from rule-based heuristics to data-driven perceptual inference, enhancing realism and user trust in immersive applications such as virtual reality, remote operation, or surgical training.

Moreover, multi-dimensional input streams enable the development of closed-loop feedback systems, where the output of the actuator dynamically adjusts in response to the user's physical interaction. For example, if slip is detected via a combination of skin stretch due to tangential force and high-frequency vibration patterns, the system can instantaneously increase grip force or modulate actuation signal to maintain contact stability. This kind of real-time responsiveness is only achievable when hardware delivers dense, temporally resolved, and perceptually relevant input.

At the system-level, rich encoded tactile data facilitate sensor fusion and control coordination across multiple modules—whether that means synchronizing the output of several fingers in a global interface, integrating haptic feedback into robotic manipulators, or coordinating vision-tactile cues in augmented reality systems. Because the data are structured in the same sensory space as human touch (e.g., friction, pressure, thermal flow), high-level software can operate at a level of abstraction that closely mirrors natural interaction. This alignment between biophysical input and psychophysical response unlocks the potential for generalized predictive models and shared feature representations across devices and platforms. Thereby paving the way for interoperable haptic systems and scalable architectures where new hardware modules or user profiles can be incorporated without retraining or redesigning core feedback algorithms.

4.6 Conclusion

In summary, by ensuring that the hardware speaks the "language" of the nervous system, the data it generates becomes meaningful to the human brain and also to the software—enabling an integrated haptic ecosystem where sensing, actuation, and cognition are tightly coupled. This co-evolution of materials, mechanics, and algorithms is what will ultimately enable haptic technologies that feel not just realistic, but indistinguishably real. To fully realize this vision, the hardware alone is not enough. The psychophysically rich signals it collects—multi-dimensional, time-varying, and context-dependent—must be interpreted, modeled, and acted upon in ways that align with human perception and behavior. This is where software plays a pivotal role.

In the next chapter, we will explore how these perceptually grounded signals can enhance software development, including how they inform predictive models of touch, adaptive feedback control, perceptual encoding strategies, and human-in-the-loop systems. By building software that not only understands the language of tactile data but also speaks it fluently, we move closer to integrated haptic systems that respond to users with the same nuance and intelligence as the human sense of touch itself.

References

1. Ahmadi, R., Arbatani, S., Ozhikandathil, J., Packirisamy, M., Dargahi, J.: A multi-purpose optical microsystem for static and dynamic tactile sensing. Sensors Actuators A Phys. **235**, 37–47 (2015)
2. Ambrosi, G., Bicchi, A., De Rossi, D., Scilingo, E.P.: The role of contact area spread rate in haptic discrimination of softness. In: Proceedings 1999 IEEE International Conference on Robotics and Automation (Cat. No. 99CH36288C), vol. 1, pp. 305–310. IEEE (1999)
3. Azenkot, S., Ladner, R.E., Wobbrock, J.O.: Smartphone haptic feedback for nonvisual wayfinding. In: The Proceedings of the 13th International ACM SIGACCESS Conference on Computers and Accessibility, pp. 281–282 (2011)
4. Bicchi, A., Scilingo, E.P., De Rossi, D.: Haptic discrimination of softness in teleoperation: the role of the contact area spread rate. IEEE Trans. Robot. Automation **16**(5), 496–504 (2000)
5. Brewster, S., Chohan, F., Brown, L.: Tactile feedback for mobile interactions. In: Proceedings of the SIGCHI Conference on Human Factors in Computing Systems, pp. 159–162 (2007)
6. Carter, T., Seah, S.A., Long, B., Drinkwater, B., Subramanian, S.: Ultrahaptics: multi-point mid-air haptic feedback for touch surfaces. In: Proceedings of the 26th Annual ACM Symposium on User Interface Software and Technology, pp. 505–514 (2013)
7. Chen, E., Marcus, B.: Force feedback for surgical simulation. Proc. IEEE **86**(3), 524–530 (1998)

8. Dhong, C., Miller, R., Root, N.B., Gupta, S., Kayser, L.V., Carpenter, C.W., Loh, K.J., Ramachandran, V.S., Lipomi, D.J.: Role of indentation depth and contact area on human perception of softness for haptic interfaces. Sci. Adv. **5**(8), eaaw8845 (2019)
9. Dobrzynska, J.A., Gijs, M.: Polymer-based flexible capacitive sensor for three-axial force measurements. J. Micromech. Microeng. **23**(1), 015009 (2012)
10. Escobar-Castillejos, D., Noguez, J., Neri, L., Magana, A., Benes, B.: A review of simulators with haptic devices for medical training. J. Med. Syst. **40**(4), 104 (2016)
11. Frediani, G., Carpi, F.: How to make the skin contact area controllable by optical calibration in wearable tactile displays of softness. Sensors **24**(20), 6770 (2024)
12. Freyberger, F.K., Färber, B.: Compliance discrimination of deformable objects by squeezing with one and two fingers. In: Proceedings of the EuroHaptics, vol. 2006, pp. 271–276 (2006)
13. Hauser, S.C., Gerling, G.J.: Force-rate cues reduce object deformation necessary to discriminate compliances harder than the skin. IEEE Trans. Hap. **11**(2), 232–240 (2017)
14. Huang, Y., Zhou, J., Ke, P., Guo, X., Yiu, C.K., Yao, K., Cai, S., Li, D., Zhou, Y., Li, J., et al.: A skin-integrated multimodal haptic interface for immersive tactile feedback. Nature Electron. **6**(12), 1020–1031 (2023)
15. Hudson, K.M., Condon, M., Ackerley, R., McGlone, F., Olausson, H., Macefield, V.G., Birznieks, I.: Effects of changing skin mechanics on the differential sensitivity to surface compliance by tactile afferents in the human finger pad. J. Neurophysiol. **114**(4), 2249–2257 (2015)
16. Jackson, B., Rosenberg, L.: Force feedback and medical simulation. In: Morgan et al. (Eds.) Interactive Technology and the New Paradigm for Healthcare, pp. 147–151. IOS Press, Amsterdam (1995)
17. Jadhav, S., Kannanda, V., Kang, B., Tolley, M.T., Schulze, J.P.: Soft robotic glove for kinesthetic haptic feedback in virtual reality environments. Electron. Imaging **29**, 19–24 (2017)
18. Jewitt, C.: A multimodal perspective on touch, communication, and learning. In: Serafini, F., Gee, E. (eds.) Remixing Multiliteracies: Theory and Practice from New London to New Times, pp. 106–118. Teachers College Press, New York (2017)
19. Jin, H., Kim, Y., Youm, W., Min, Y., Seo, S., Lim, C., Hong, C.H., Kwon, S., Park, G., Park, S., et al.: Highly pixelated, untethered tactile interfaces for an ultra-flexible on-skin telehaptic system. npj Flexible Electron. **6**(1), 82 (2022)
20. Koskinen, E., Kaaresoja, T., Laitinen, P.: Feel-good touch: finding the most pleasant tactile feedback for a mobile touch screen button. In: Proceedings of the 10th International Conference on Multimodal Interfaces, pp. 297–304 (2008)
21. Lambeta, M., Chou, P.W., Tian, S., Yang, B., Maloon, B., Most, V.R., Stroud, D., Santos, R., Byagowi, A., Kammerer, G., et al.: Digit: A novel design for a low-cost compact high-resolution tactile sensor with application to in-hand manipulation. IEEE Robot. Automation Lett. **5**(3), 3838–3845 (2020)
22. LaMotte, R.H.: Softness discrimination with a tool. J. Neurophysiol. **83**(4), 1777–1786 (2000)
23. Lawrence, D.A., Pao, L.Y., Dougherty, A.M., Salada, M.A., Pavlou, Y.: Rate-hardness: A new performance metric for haptic interfaces. IEEE Trans. Robot. Automation **16**(4), 357–371 (2000)
24. Laycock, S.D., Day, A.: Recent developments and applications of haptic devices. In: Computer Graphics Forum, vol. 22, pp. 117–132. Wiley Online Library, Hoboken (2003)
25. Lee, M.A., Zhu, Y., Zachares, P., Tan, M., Srinivasan, K., Savarese, S., Fei-Fei, L., Garg, A., Bohg, J.: Making sense of vision and touch: learning multimodal representations for contact-rich tasks. IEEE Trans. Robot. **36**(3), 582–596 (2020)
26. Lepora, N.F., Ward-Cherrier, B.: Superresolution with an optical tactile sensor. In: 2015 IEEE/RSJ International Conference on Intelligent Robots and Systems (IROS), pp. 2686–2691. IEEE (2015)
27. Lezkan, A., Metzger, A., Drewing, K.: Active haptic exploration of softness: Indentation force is systematically related to prediction, sensation and motivation. Front. Integr. Neurosci. **12**, 59 (2018)

28. Li, B., Gerling, G.J.: Individual differences impacting skin deformation and tactile discrimination with compliant elastic surfaces. In: 2021 IEEE World Haptics Conference (WHC), pp. 721–726. IEEE (2021)
29. Li, B., Gerling, G.J.: An individual's skin stiffness predicts their tactile discrimination of compliance. J. Physiol. **601**(24), 5777–5794 (2023)
30. Li, B., Hauser, S., Gerling, G.J.: Identifying 3-d spatiotemporal skin deformation cues evoked in interacting with compliant elastic surfaces. In: 2020 IEEE Haptics Symposium (HAPTICS), pp. 35–40. IEEE (2020)
31. Li, B., Lee, J., Gerling, G.J.: A magnetorheological elastomer device for programmable actuation and sensing of soft haptic experiences. In: 2021 IEEE World Haptics Conference (WHC), pp. 349–349. IEEE (2021)
32. Li, B., Hauser, S.C., Gerling, G.J.: Faster indentation influences skin deformation to reduce tactile discriminability of compliant objects. IEEE Trans. Hap. **16**(2), 215–227 (2023)
33. Li, B., Lee, J., Gerling, G.J.: Perceptibility of programmable softness displays using magnetorheological elastomers. In: Proceedings of the Human Factors and Ergonomics Society Annual Meeting, vol. 67, pp. 1518–1524. SAGE Publications, Los Angeles (2023)
34. Lin, S., Zhu, L., Qiu, Y., Jiang, Z., Wang, Y., Zhu, J., Wu, H.: A self-powered multi-functional sensor based on triboelectric nanogenerator for monitoring states of rotating motion. Nano Energy **83**, 105857 (2021)
35. Nishino, H., Goto, R., Kagawa, T., Yoshida, K., Utsumiya, K., Hirooka, J., Osada, T., Nagatomo, N., Aoki, E.: A touch screen interface design with tactile feedback. In: 2011 International Conference on Complex, Intelligent, and Software Intensive Systems, pp. 53–60. IEEE (2011)
36. Noh, Y., Bimbo, J., Sareh, S., Wurdemann, H., Fraś, J., Chathuranga, D.S., Liu, H., Housden, J., Althoefer, K., Rhode, K.: Multi-axis force/torque sensor based on simply-supported beam and optoelectronics. Sensors **16**(11), 1936 (2016)
37. Perret, J., Vander Poorten, E.: Touching virtual reality: a review of haptic gloves. In: ACTUATOR 2018; 16th International Conference on New Actuators, pp. 1–5. VDE (2018)
38. Ruspini, D.C., Kolarov, K., Khatib, O.: Haptic interaction in virtual environments. In: Proceedings of the 1997 IEEE/RSJ International Conference on Intelligent Robot and Systems. Innovative Robotics for Real-World Applications. IROS'97, vol. 1, pp. 128–133. IEEE (1997)
39. Salcudean, S.E.: Control for teleoperation and haptic interfaces. In: Control Problems in Robotics and Automation, pp. 51–66. Springer, Berlin (2005)
40. Schorr, S.B., Okamura, A.M.: Three-dimensional skin deformation as force substitution: Wearable device design and performance during haptic exploration of virtual environments. IEEE Trans. Hap. **10**(3), 418–430 (2017)
41. Sferrazza, C., D'Andrea, R.: Design, motivation and evaluation of a full-resolution optical tactile sensor. Sensors **19**(4), 928 (2019)
42. Shi, Y., Shen, G.: Haptic sensing and feedback techniques toward virtual reality. Research **7**, 0333 (2024)
43. Shi, J., Dai, Y., Cheng, Y., Xie, S., Li, G., Liu, Y., Wang, J., Zhang, R., Bai, N., Cai, M., et al.: Embedment of sensing elements for robust, highly sensitive, and cross-talk–free iontronic skins for robotics applications. Sci. Adv. **9**(9), eadf8831 (2023)
44. Sinclair, M., Ofek, E., Gonzalez-Franco, M., Holz, C.: Capstancrunch: A haptic vr controller with user-supplied force feedback. In: Proceedings of the 32nd Annual ACM symposium on User Interface Software and Technology, pp. 815–829 (2019)
45. Sonar, H.A., Gerratt, A.P., Lacour, S.P., Paik, J.: Closed-loop haptic feedback control using a self-sensing soft pneumatic actuator skin. Soft Robot. **7**(1), 22–29 (2020)
46. Song, A., Wu, J., Qin, G., Huang, W.: A novel self-decoupled four degree-of-freedom wrist force/torque sensor. Measurement **40**(9–10), 883–891 (2007)
47. Srinivasan, M.A., LaMotte, R.H.: Tactual discrimination of softness. J. Neurophysiol. **73**(1), 88–101 (1995)

48. Stanley, A.A., Hata, K., Okamura, A.M.: Closed-loop shape control of a haptic jamming deformable surface. In: 2016 IEEE International Conference on Robotics and Automation (ICRA), pp. 2718–2724. IEEE (2016)
49. Stone, R.J.: Haptic feedback: A brief history from telepresence to virtual reality. In: International Workshop on Haptic Human-Computer Interaction, pp. 1–16. Springer (2000)
50. Sziebig, G., Solvang, B., Kiss, C., Korondi, P.: Vibro-tactile feedback for vr systems. In: 2009 2nd Conference on Human System Interactions, pp. 406–410. IEEE (2009)
51. Takahashi, H., Nakai, A., Thanh-Vinh, N., Matsumoto, K., Shimoyama, I.: A triaxial tactile sensor without crosstalk using pairs of piezoresistive beams with sidewall doping. Sensors Actuators A Phys. **199**, 43–48 (2013)
52. Tan, H.Z., Durlach, N.I., Shao, Y., Wei, M.: Manual resolution of compliance when work and force cues are minimized. ASME Dyn. Syst. Control Div. Publ. DSC ASME, New York, NY (USA) **49**, 99–104 (1993)
53. Tan, H.Z., Durlach, N.I., Beauregard, G.L., Srinivasan, M.A.: Manual discrimination of compliance using active pinch grasp: The roles of force and work cues. Percept. Psychophys. **57**(4), 495–510 (1995)
54. Tanaka, Y., Bergmann Tiest, W.M., Kappers, A.M., Sano, A.: Contact force and scanning velocity during active roughness perception. PloS One **9**(3), e93363 (2014)
55. Tiest, W.M.B., Kappers, A.M.: Cues for haptic perception of compliance. IEEE Trans. Hap. **2**(4), 189–199 (2009)
56. Vega-Bermudez, F., Johnson, K.O.: Fingertip skin conformance accounts, in part, for differences in tactile spatial acuity in young subjects, but not for the decline in spatial acuity with aging. Percept. Psychophys. **66**(1), 60–67 (2004)
57. Wang, R., Hu, S., Zhu, W., Huang, Y., Wang, W., Li, Y., Yang, Y., Yu, J., Deng, Y.: Recent progress in high-resolution tactile sensor array: From sensor fabrication to advanced applications. Progress Natural Sci. Mater. Int. **33**(1), 55–66 (2023)
58. Welcome, D.E., Dong, R.G., Xu, X.S., Warren, C., McDowell, T.W.: The effects of vibration-reducing gloves on finger vibration. Int. J. Ind. Ergon. **44**(1), 45–59 (2014)
59. Xu, C., Hauser, S.C., Wang, Y., Gerling, G.J.: Roles of force cues and proprioceptive joint angles in active exploration of compliant objects. In: 2019 IEEE World Haptics Conference (WHC), pp. 353–358. IEEE (2019)
60. Yao, N., Wang, S.: Recent progress of optical tactile sensors: A review. Optics Laser Technol. **176**, 111040 (2024)
61. Yuan, W., Dong, S., Adelson, E.H.: Gelsight: High-resolution robot tactile sensors for estimating geometry and force. Sensors **17**(12), 2762 (2017)

Chapter 5
Software as Pathways for Perception

Abstract Just as raw nerve signals become perception when processed by the brain, tactile data only become experience when decoded through models that reflect the logic of human perception. This is where software serves the engine of haptic intelligence: by inferring what the user feels, not just what the sensor detects, software bridges the mechanical interface and perceptual reality. By building psychophysical inference models that operate directly on afferent-level inputs, the software subsystem becomes capable of predicting what a user feels, selecting the optimal combination of cues for a given interaction, and continuously adapting in real-time. This marks a fundamental shift in haptics from controlling devices to understanding touch, paving the way for integrated systems that deliver sensations with the nuance, adaptability, and meaning of natural human touch.

5.1 Functional Role of Software Subsystems in Inference

In an integrated haptic system, the software subsystem plays a critical intermediary role, forming the conceptual and functional bridge between the hardware subsystem, which senses and digitizes physical contact events, and the application subsystem, which operates at the level of perceptual meaning, user interaction, and system-level decision-making.

On one end, the hardware subsystem comprises tactile sensors, transducers, and signal circuits that capture raw physical variables such as normal force, shear stress, temperature, vibration, and surface deformation. These signals are rich in physical detail but lack interpretability on their own, especially in the context of how humans perceive and respond to touch.

On the other end, the application subsystem must operate on interpretable, task-relevant representations that map onto perceptual constructs such as softness, slipperiness, warmth, texture, or intent. For instance, a robotic gripper must decide whether to adjust its grip force based on whether an object feels compliant or slippery, not merely on changes in voltage or sensor resistance.

It is within this gap—the space between measured physical quantities and perceptually meaningful interpretations—that the software subsystem operates. Its

B. Li, T. Cody, *Design Principles for Integrated Haptic Systems*, Springer Series on Touch and Haptic Systems, https://doi.org/10.1007/978-3-032-16986-0_5

fundamental function is to transform low-level sensor data into high-level perceptual information, enabling the system to reason about tactile events in a way that aligns with human psychophysics and task demands. In doing so, the software subsystem performs perceptual inference: extracting informative features, weighting them by context, and synthesizing them into coherent perceptual states.

As in the previous chapter, the software subsystem operates under the same two foundational principles of haptic design. The first principle—that haptic capability is defined by the scope and use of cues—extends here into the informational domain: software determines which sensed variables enter perceptual computation and how they are combined to represent tactile events. The second principle—that the correlation between cues and perception varies between individuals—emerges as the system's requirement for adaptive inference, enabling perceptual meaning to adjust across users, contexts, and goals. Through these principles, the software subsystem does not merely pass information between hardware and application; it governs the transformation of measurement into perception, ensuring that haptic information remains both physically grounded and psychophysically valid.

This need for transformation makes inference the dominant framing of the software subsystem in integrated haptic systems. Unlike stimulus-driven architectures that treat signals as fixed mappings to sensations, inference-centric systems view perception as a process of reasoning under uncertainty—predicting how a given mechanical interaction will be felt, not just how it was measured. The software subsystem thus becomes a perceptual engine—extracting meaningful structure from high-dimensional sensor data, weighting cues by context, and continuously updating its internal models to minimize perceptual error. In this view, intelligence in haptics arises not from reproducing stimuli but from understanding them: a shift from rendering touch to reasoning about it.

5.2 Biological Analogues for Software Subsystems

Biology clarifies why this inference framing is necessary. The human nervous system does not passively react to sensory inputs; it predicts, tests, and corrects them against internal models of the world [1, 24–26, 30]. The resulting percept is therefore an inference: a best estimate of the external cause of sensory data given prior experience and context. To emulate this intelligence, haptic systems must evolve from reproducing physical signals to predicting how those signals will be felt. Sensors thus provide not only mechanical measurements but also a high-dimensional state of interaction comprised of patterns of force, deformation, and vibration that mirror the neural codes of biological skin. The software then functions as an inference engine, mapping these mechanical states to perceptual outcomes and adjusting actuation to minimize perceptual error, ensuring that the user feels what the system intends, even under variable conditions.

Achieving such perceptual intelligence requires a shift from rule-based control toward data-driven models that learn how humans perceive tactile stimuli from rich, multimodal data. Traditional haptic systems compress this complexity into a few scalar cues—typically normal force and vibration—then replay them in an open-loop, hoping to elicit the intended sensation. Such heuristics cannot adapt to individual differences, material variation, or new interaction contexts. In contrast, inference-based systems employ predictive models of psychophysical response trained on data that capture the biomechanical and perceptual coupling of touch. These models translate sensor-level information into latent perceptual variables that can be used for feedback and control.

5.2.1 Leveraging Afferent-Level Signals

Psychophysical inference becomes neurophysiologically valid only when the model's input mirrors the actual information delivered to the brain by peripheral afferents. In human somatosensory system, first-order neurons do not perform categorical abstraction or pre-averaging; instead, they transmit rich, millisecond-scale patterns of receptor-specific discharges that jointly encode normal pressure, tangential slips, high-frequency vibrations, and temperature gradients. To replicate this encoding fidelity, the psychophysical inference models should operate directly on the raw, synchronized outputs of advanced haptic transducers of mechanical stimuli into meaningful information, preserving this "afferent-level" complexity rather than reducing input to engineered descriptors (e.g., root-mean-square force or mean vibration amplitude).

Preserving the full mechanical state vector (i.e., all the tactile-relevant biophysical variables that together define the state of contact) serves two critical purposes. First, it allows the learning algorithm to discover nonlinear, cross-modality interactions—such as the coincidence of shear-rate spikes and high-frequency bursts at slip onset—that are known to drive fast adapting mechanoreceptor firing in the human system but can be made invisible once signals are pre-processed. Second, it ensures that the model's internal representations remain structurally aligned with biological coding. As a result, when the model infers perceptual dimensions, such as compliance, roughness, or friction, it does so by weighting feature channels in ways that are interpretable through the lens of established mechanoreceptor physiology.

The neurophysiological organization of the human tactile system demonstrates that afferent-level representations are not only sufficient for perception but also foundational for adaptability. By preserving the full dimensionality of these inputs, the human nervous system maintains a general-purpose substrate for inference: one that can be dynamically reweighted to meet the demands of new tasks, materials, or contexts. This biological adaptability provides the design motivation for integrated haptic systems whose predictive models are themselves neurophysiologically valid: by aligning with the structures that support general-purpose tactile inference in humans, such systems can inherit the flexibility and efficiency of natural touch.

5.2.2 *Adaptive Cue Selection and Re-Weighting*

The brain does not rely on a fixed set of tactile cues for all tasks; rather, it dynamically selects and re-weights incoming signals based on stimulus properties and behavioral goals. This flexibility allows the somatosensory system to optimize perceptual accuracy across diverse contexts—such as distinguishing surface compliance versus detecting edge orientation. Psychophysical models grounded in biomechanically rich inputs can inherit this capacity for adaptive cue selection, making them not only perceptually valid but also perceptually intelligent.

For example, studies of compliance perception have demonstrated that the optimal combination of biomechanical cues varies with the mechanical properties of the stimulus. When interacting with soft stimuli (e.g., Young's modulus ≈ 33 kPa), compliance judgments were best explained by a combination of surface curvature and contact area. However, for stiffer stimuli (e.g., ≈ 184 kPa), the discriminative power shifted to normal contact force and contact area [12, 15, 17].

Just as the human brain selectively attends to different tactile cues depending on the interaction goal [5, 10, 13, 29], psychophysical inference models can exploit this structure by learning which subsets of features are most relevant under different interaction conditions. Trained on high-dimensional, high-spatial-temporal data, such models can build internal representations of tasks and materials by identifying and prioritizing the most informative cues. During real-time operation, they can assign higher perceptual weights to the most relevant channels—favoring contact area and curvature cues during compliance discrimination for soft stimuli, vibration spectra during fine texture sensing, or shear-rate during slip detection. In doing so, the models mirror the human brain's strategy of task- and context-sensitive cue optimization.

In sum, psychophysical inference models mirror the adaptive intelligence of the human sensory system by achieving three key capabilities: optimal cue selection to reduce redundancy, dynamic cue-reweighting to meet task-specific demands, and personalized calibration based on individual differences. These functions allow the model to prioritize the most perceptually informative signals, adapt its strategy in real-time, and deliver tailored tactile feedback for each user—just as the human brain does through efficient coding, top-down modulation, and sensory learning, described below and as Table 5.1 shows.

5.2.2.1 Redundancy Reduction for Efficient Tactile Coding

This task-dependent cue selection and prioritization also enables psychophysical inference models to minimize cue redundancy—mirroring strategies employed by the human sensory system [4, 7]. In natural touch, the nervous system dynamically down-weights cues that no longer offer unique information, optimizing perceptual inference. Neurophysiological studies suggest that the central nervous system performs efficient coding by filtering overlapping inputs to reduce cognitive and metabolic load while preserving perceptual accuracy. For example, neurons in the

Table 5.1 Human versus model-based strategies for efficient, context-sensitive, and personalized tactile inference

Capability	Human sensory system	Psychophysical inference model
Optimal cue selection	Neural circuits de-correlate afferent inputs, prioritizing novel signals	Data-driven pruning of correlated features for compact representation
Task-dependent cue re-weighting	Top-down modulation alters afferent gains for the current task	Context-conditioned feature weights emphasizing the most relevant sensor channels
Personalized calibration	Experience tunes synaptic weights to each individual's physical characteristics	Calibration routines adjust feature weights according to the user's sensitivity profile

somatosensory cortex often exhibit sparse coding—compressing high-dimensional sensory inputs into a reduced set of highly informative activity patterns. Additionally, sensory cortices adapt to continuous stimulation by dynamically retuning receptive fields, enhancing responsiveness to novel or task-relevant features while attenuating predictable or redundant ones [9, 21, 28].

This principle of redundancy reduction has been observed in empirical studies of tactile compliance perception. In a study using stereo imaging to reconstruct 3D skin deformation, researchers analyzed five biomechanical cues—contact area, curvature, indentation depth, eccentricity, and contact force—across materials of varying compliance [14, 17]. Statistical comparisons using t-tests and correlation matrices revealed that these cues were only weakly correlated, with no significant multicollinearity, confirming that each cue contributed non-overlapping, perceptually relevant information for compliance discrimination. Such statistical independence supports the hypothesis that human perception of compliance relies on a multi-cue integration strategy in which redundant signals are minimized.

For psychophysical inference models trained on rich, synchronized biomechanical data, this selective pruning and retuning becomes functionally valuable. It allows the model to discard irrelevant or redundant channels, reduce dimensionality, and emphasize those features that best align with the perceptual tasks. This biologically inspired coding scheme produces compact, efficient, and behaviorally meaningful representations—enabling faster, more accurate perceptual inference while minimizing computational overhead.

5.2.2.2 Re-Weighting Based on Tasks

In natural touch, the human brain does not treat all tactile cues equally across every situation—it dynamically adjusts the perceptual gain of incoming afferent channels depending on the interaction goal. This flexible re-weighting strategy is thought to emerge from top-down cortical modulation, which fine-tunes the sensitivity of afferent pathways to optimize perception for a given task [3, 20, 22]. For example, when exploring a textured surface, the brain prioritizes temporal vibrational patterns conveyed by fast-adapting afferents; when assessing compliance levels, it shifts attention towards cues derived from skin deformation captured by SA-I and SA-II afferents.

Psychophysical models trained on high-dimensional biomechanical data can mirror this capacity for dynamic cue re-weighting. Instead of treating cue importance as fixed, these models learn to adjust their internal weighting of cues based on the task context. For instance, through supervised learning (using labeled perceptual judgments) or reinforcement learning (driven by task success), the model can discover which input features are most informative for different perceptual goals. When detecting incipient slip, for example, the model may up-weight fast-changing shear rates and high-frequency vibratory spikes; when estimating shape or edges, it may prioritize the local skin stretch and the change rate of normal force. This adaptive re-weighting allows haptic systems to intelligently modulate the feedback they provide, delivering only the most relevant mechanical cues for the intended goal at hand. Such task-optimized encoding improves perceptual clarity, reduces extra data processing, and lowers cognitive load.

5.2.2.3 Calibration to Users

In addition to task-specific tuning, the human tactile system also adapts to long-term individual differences in skin mechanics and sensory physiology. People vary significantly in tactile sensitivity due to age, skin properties (e.g., stiffness, hydration, thickness), fingerprint ridge density, and neural functioning—all of which affect how mechanical stimuli are transduced into perception [2, 8, 15, 23]. The nervous system calibrates to these differences over time by implicitly learning a unique internal mapping between mechanical input and perceptual meaning for each individual. Psychophysical inference models, when built from biologically and perceptually aligned data, can emulate this personalized adaptation by incorporating user-specific calibration data—whether from pretests, in-task behavioral feedback, or inferred psychophysical thresholds—the model can re-weight input cues to reflect how a specific user interprets touch. For example, if a user's skin is stiffer than average, leading to reduced sensitivity to skin deformation, then the model can down-weight deformation-derived cues and amplify force-based ones. Over time, this process forms a personalized perceptual profile that tailors the feedback to each user's sensory characteristics—without changing the hardware. Such user-specific adaptation makes haptic devices more inclusive, consistent, and accessible across populations with different tactile abilities, including the elderly or individuals with sensory impairments.

5.3 Multimodal Integration and Perceptual Coherence

The integration of sensing and control in haptic systems is not simply an engineering necessity—it is a structural consequence of the nature of touch itself. As discussed in

Chap. 4, human touch is inherently multichannel and multimodal: each contact event simultaneously excites multiple receptor types, each tuned to distinct mechanical and temporal features. Perception emerges not from any single cue, but from the brain's ability to integrate these heterogeneous signals into coherent, high-level judgments such as compliance, texture, or slipperiness. Inference-centric haptic systems inherit this property by design. Once sensor data are treated as afferent-level inputs to a perceptual model, integration is no longer a matter of combining signals; it becomes the process through which perception is constructed.

The demand for coordination follows directly. A system that reasons about perception rather than merely measuring signals must synchronize sensing and actuation so that predictions, feedback, and perceptual outcomes remain aligned. As haptic systems incorporate multiple sensing modalities (e.g., force, vibration, temperature) and distributed hardware elements (e.g., multi-DoF actuators, tactile arrays), their stability and coherence depend on the principled management of these coupled data streams. Without such coordination, the system cannot preserve the perceptual unity that multimodal integration in biology naturally maintains. Hence, sensor integration and control coordination represent not just implementation challenges, but the necessary architectural expression of an inference-centric, neurophysiologically grounded design philosophy.

5.3.1 Design Implications for Multimodal Integration

In an inference-centric framework, sensor integration is the computational analogue of multisensory convergence: to achieve perceptual validity, multimodal streams must be fused with temporal and spatial precision so that inference operates on a unified perceptual substrate. Conventional systems often rely on hard-coded fusion rules or independent processing pipelines. These approaches treat each data modality as an independent signal, and coordination among control outputs is typically heuristic. Such methods lack robustness to signal variability, noise, and user differences, and they often fail to prioritize signals that matter most for perception. As a result, these systems cannot reason about user experience or dynamically adapt to changing conditions.

Data-driven psychophysical inference models offer a more principled alternative. Trained on high-dimensional, perceptually relevant sensor inputs, these models learn the relationship between "afferent-level" cues and perceptual outcomes, performing sensor fusion in "perceptual space" rather than "signal space". Key capabilities of psychophysical inference models for sensor fusion include task-dependent weighting, where the system prioritizes the most perceptually informative modalities based on the interaction goal. Additionally, they support user-specific fusion strategies, adapting weights based on individualized calibration data to reflect differences in tactile sensitivity due to factors like skin stiffness or age. They enable distributed coherence, integrating sensor inputs across multiple contact points—

such as multiple fingers engaging with different surfaces—into a unified, high-level percept.

Integrating information channels from diverse, multimodal sensors and hardware components goes beyond combining data—it is about preserving the perceptual logic of the nervous system. In this sense, sensor fusion becomes an act of perceptual reconstruction: by aligning the informational structure of artificial systems with that of biology, these models deliver coherent, context-sensitive feedback that remains robust to noise, sensor drift, and inter-user variability. Key considerations for integrating diverse sensor data streams in psychophysical inference models include:

Temporal Synchronization Temporal alignment is fundamental. In the human nervous system, perceptual meaning often arises from the precise timing of multimodal signals—for instance, the co-occurrence of lateral skin stretch (shear rate) and high-frequency vibrations at slip onset. To preserve these structures, all sensor outputs must be synchronized to a common clock, ideally with sub-millisecond precision that is close to human-level sensitivity. Temporal misalignment can cause cross-modal de-correlation, weakening the model's ability to learn true perceptual mappings and forcing reliance on heuristics.

Spatial Co-registration Signals must not only be temporally aligned, but also spatially co-registered to the same contact region. This means that transducers capturing different functions (e.g., normal vs tangential force) or modalities (e.g., vibration, temperature) must be either collocated in a global coordinate system or with minimal offsets. Inconsistencies in spatial registration introduce artifacts when integrating cues (e.g., skin stretch measured 2 millimeters away from where force is applied can mislead a compliance-prediction model).

Sampling Rate and Bandwidth Matching Each sensor must operate at a sampling rate appropriate for the perceptual channel it targets. For example, in reference to Chap. 2, vibration sensors (targeting FA afferents) should sample at > 1 kilohertz to resolve high-frequency bursts. Force sensors may operate at lower rates, but must still synchronize with high-rate signals to align dynamics. Failure to match perceptually relevant bandwidths leads to signal aliasing or under-representation of critical cues.

Noise Profiles and Cross-Modal Filtering Sensor fusion must account for the heterogeneous noise characteristics of different modalities. Accelerometers, for example, are sensitive to motion artifacts, while strain gauges may be subject to baseline drift. Fusion algorithms must dynamically weigh inputs based on signal reliability, often using signal-to-noise ratio or real-time confidence metrics.

Latency Uniformity Sensors must also be latency-matched so that all modalities arrive at the fusion point without temporal lag distortion. If one sensor (e.g., force) has lower latency than another (e.g., surface deformation from a camera), the model must delay the fast input, interpolate the slow one to maintain coherence, or operate with incomplete information.

Standardized Data Interface Furthermore, to ensure plug-and-play accessibility and minimize integration errors, transducers should conform to standardized data formats and communication protocols, which are ideally equipped with time-stamped packets, unit normalization, and metadata describing spatial location, modality, and calibration status.

5.4 Predictive Modeling of Compliance Perception

The software subsystem bridges the gap between the raw mechanical signals measured by the hardware and the perceptually meaningful information required by the application. Its functional role is to extract, organize, and infer, transforming streams of physical tactile data into high-level representations that align with human perceptual and functional percepts such as softness or slipperiness. In doing so, the software subsystem enables the system not merely to measure contact events but to understand them.

This capability is exemplified through the case study of compliance perception. The hardware subsystem captures a set of informative deformation cues, including contact area, curvature, penetration depth, eccentricity, and force, that jointly characterize the mechanical interactions between the fingertip and the stimulus surface. The software subsystem then operates on these cues to construct predictive models capable of estimating psychophysical responses across individuals with differing finger size, stiffness, and ridge breadth. By learning which deformation features most strongly predict perceived compliance, the software layer effectively quantifies the perceptual logic embedded in the sensor data. Capturing these patterns allows the model to forecast each individual's discrimination performance, revealing how tactile perception can be computationally inferred from biomechanical measurements.

Moreover, the software subsystem's inferential scope extends beyond predicting perception. When certain cues are not directly measurable, due to sensing constraints or design choices, the software can predict these unobserved biomechanical variables from the available data. For instance, as shown later in this chapter, contact force can be accurately estimated from a combination of geometric deformation cues such as area, curvature, and depth.

5.4.1 Predicting Perceptual Responses from Deformation Cues

To demonstrate the software subsystem's ability to align tactile data with perceptual meaning, we first examine how predictive models can be used to estimate compliance perception across individuals. Compliance, arises from complex interactions between the stimulus and the fingertip's biomechanical response. Because these interactions vary across individuals with different skin properties and physical characteristics, perceptual outcomes are inherently personalized.

5.4.1.1 Clustering Based on Skin Properties

By leveraging informative skin deformation cues captured by the hardware, such as contact area, curvature, penetration depth, eccentricity, and force, the software subsystem can construct models that predict how each individual discriminates between compliant stimuli. In an existing study, participants were first characterized based on the physical characteristics of their fingers—size, stiffness, and ridge breadth—using K-means clustering [15, 27], see Fig. 5.1. Two distinct cohorts emerged. One group exhibited smaller, more compliant (softer) skin; the other, bigger and less compliant (stiffer) fingers, as discussed in Chap. 3. This division reflects underlying differences in how each group's skin responded to indentation in passive touch, suggesting that perceptual outcomes may be shaped by the mechanical events occurred on the skin. Moreover, this clustering allowed for group-specific modeling of perceptual inference, acknowledging that individuals do not rely on the same cues, nor in the same way, when performing compliance discrimination tasks [14, 16].

5.4.1.2 Informativeness of Deformation Cues

Once these cohorts were established, the next step was to determine which aspects of skin deformation carried the most perceptual weight. To determine which cues

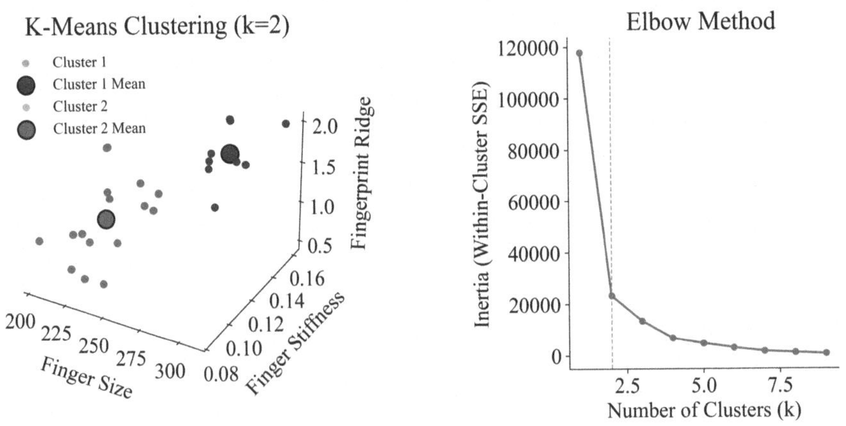

Fig. 5.1 Participant grouping obtained by K-means clustering based on finger size, finger stiffness, and ridge breadth. Two distinct cohorts emerged: Group 1 (softer fingers) and Group 2 (stiffer fingers), reflecting underlying differences in skin mechanics across individuals [15]

were most informative for each group, five skin deformation metrics were extracted during fingertip-object interaction: contact area, surface curvature, indentation depth, eccentricity, and contact force [16, 17]. Principal Component Analysis (PCA) [11, 15] was applied separately within each group to quantify the variance explained by each cue in relation to psychophysical performance.

The analysis revealed that contact area is an informative cue for both groups, consistent with its foundational role in compliance perception. Group 1 (softer fingers) relied more heavily on curvature and penetration depth, suggesting a sensitivity to geometric deformation. Group 2 (stiffer fingers) relied more on eccentricity and force, possibly due to more localized contact shape and higher resistance to penetration biomechanically. These results indicate that the relative weighting of deformation cues is varied, modulated at least by skin properties.

5.4.1.3 Building and Validating Predictive Models

These group-specific patterns form a foundation for the integrated haptic system design process for inference-centric haptic systems, which can possibly predict psychophysical responses even for new individuals or unseen conditions.

To test the viability of group-specific models, separate predictive models (e.g. multivariant linear regression models [14, 15]) were constructed for each cohort using the five deformation cues as inputs and psychophysical performance (e.g., discrimination accuracy) as output. Model generalizability was evaluated through a leave-six-participants-out (test sample size = 216) cross-validation (fivefold) scheme. That is, models were trained on the majority of participants and tested on left-out individuals to assess how well the model predicted unseen perceptual responses, see Fig. 5.2.

The results showed a high correlation between predicted and actual performance, demonstrating that software models informed by informative cues can effectively predict perceptual outcomes. Tailoring models to specific user groups (rather than

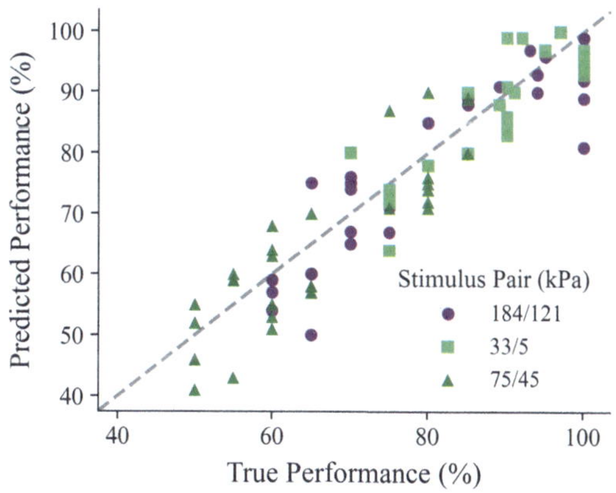

Fig. 5.2 Model validation through unseen data (n=108) prediction. Predicted compliance-discrimination performance shows high correlation with measured psychophysical results across stimulus modulus, confirming that deformation-based predictive models capture the link between skin mechanics and perception [15]

using a generic model across all users) significantly improves prediction accuracy. And, importantly, the software subsystem is capable of inferring individual perceptual responses from deformation cues alone, without requiring direct user feedback.

This illustrates a core function of the software subsystem: mapping physical inputs into perceptual outcomes, in a way that reflects both population-level patterns and individual-level variability. Such personalized perceptual modeling is essential for enabling haptic systems to adapt across users—whether in rendering appropriate feedback, customizing device behavior, or enabling user-specific closed-loop control.

5.4.2 *Predictive Models for Skin Deformation Cues*

Beyond predicting perceptual responses, the software subsystem can also infer how contact interactions behave mechanically. It can reconstruct biomechanical cues that may not be directly measurable by the hardware from the relationships embedded in available deformation data. This capability becomes crucial when practical or design constraints limit the number or type of sensors in the system. For example, a camera system can capture surface deformation patterns but lack direct measurement of normal force. In such case, the software can learn the underlying relationships among deformation features and use them to predict missing mechanical cues.

In this section, we demonstrate how the software models can predict contact force based on informative skin deformation cues such as contact area, curvature, penetration depth, and eccentricity [18]. These cues collectively encode how the skin deforms under load.

5.4.2.1 Correlations Among Skin Deformation Cues

The first step toward predictive modeling is understanding the statistical relationships between cues. See Fig. 5.3, the correlation matrix across five cues, under two stimulus conditions: a soft stimulus (45 kPa) and a slightly stiffer stimulus (75 kPa), reveal that the five deformation cues are largely independent—each encodes a distinct mechanical aspect of skin deformation—but all maintain measurable correlations with contact force. The pattern of these correlations shifts with stimulus modulus: for softer materials, contact area, penetration depth, and curvature all exhibit stronger links to force, whereas for stiffer materials, contact area and eccentricity become predictive. This variation suggests that the mapping from deformation geometry to force is context-dependent, governed by both the material being touched and the mechanics of the skin–object interface, motivating the development of specific prediction models rather than a general one.

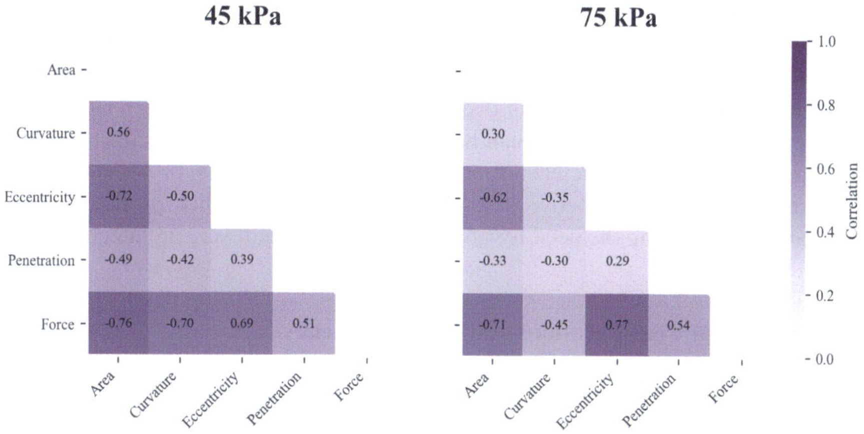

Fig. 5.3 Correlation matrices among five deformation cues (contact area, curvature, penetration depth, eccentricity, and force) under 45 and 75 kPa stimuli[18]

5.4.2.2 Cross-Model Validation

To evaluate model performance and generalizability, cross-model prediction tests were performed, see Fig. 5.4. Three regression models were trained on data from one stimulus modulus (45 kPa) and tested on either the same or a different modulus

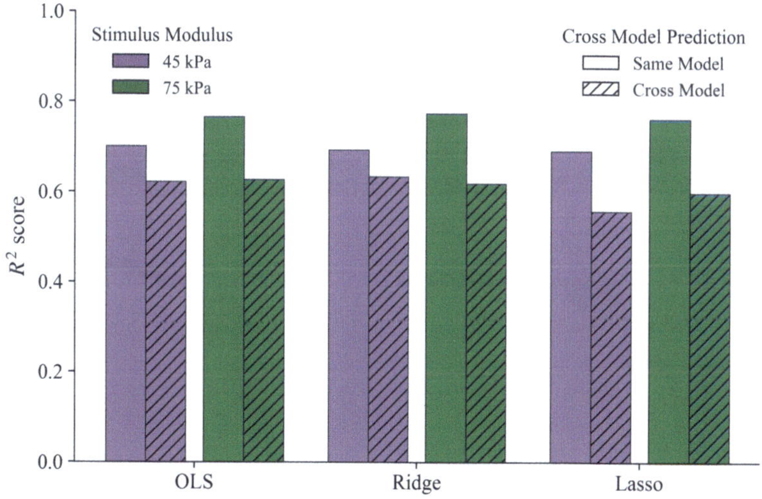

Fig. 5.4 Cross-model validation of contact-force prediction. Models trained on 45 kPa data perform well when tested on the same modulus but show reduced accuracy when applied to 75 kPa data across three regression methods, indicating the need for modulus-specific predictive models[18]

(75 kPa), and similarly applied to the other stimulus modulus (75 kPa). In the same-modulus cases, models accurately predicted contact force from the remaining four cues. However, when applied across moduli, performance declined consistently for all three regression models. This decrease indicates that while the relationships among cues are presented, the quantitative mapping is modulus-specific. A single universal model is insufficient—modulus-conditioned models produce superior predictive accuracy and better align with the physical reality of deformation mechanics, reinforcing the need for adaptive or modular software architectures.

5.4.2.3 Design Considerations and Data Sufficiency

Finally, from an design standpoint, predictive performance must be weighed against data requirements, considering the amount of data required to achieve stable prediction performance [6, 19]. See Fig. 5.5, the number of samples needed to reach a stable prediction threshold varies with stimulus modulus. Estimating contact force for a 45 kPa stimulus requires approximately 30 samples to achieve steady performance, whereas only about 20 samples suffice for a 75 kPa stimulus. This aligns with the scientific observation that stiffer stimuli produce more pronounced and consistent force-related deformation cues, enabling faster model convergence. This analysis highlights a fundamental tradeoff in software subsystem design: achieving reliable inference depends not only on cue quality and model type, but also on data sufficiency, which is in turn influenced by the physical properties of the stimuli.

These results demonstrate that the software subsystem's inferential capability extends beyond perceptual modeling to the reconstruction of latent biomechanical states. By learning context-dependent mappings among deformation features, it can compensate for sensing limitations while maintaining high predictive accuracy, embodying the inference-centric concept that intelligence in haptic systems arises

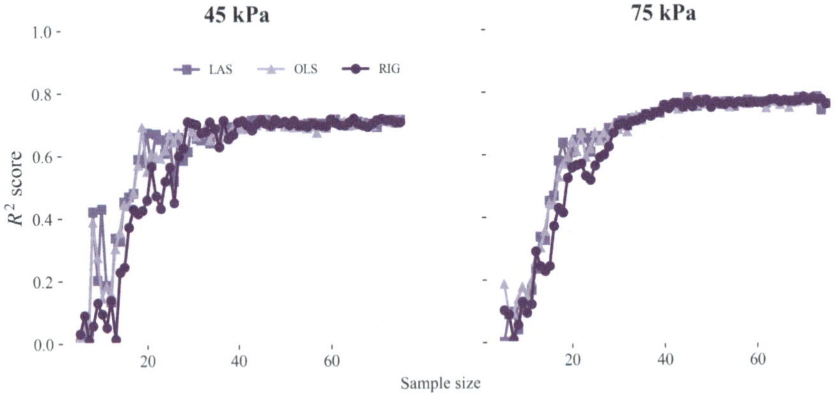

Fig. 5.5 Sample-size analysis showing the number of training samples required to achieve stable predictive performance[18]

from reasoning effectively with what is sensed, rather than measuring everything directly.

This process mirrors how biological perception compensates for sensory limitations. The human tactile system rarely measures force explicitly; instead, it infers it from patterns of skin stretch, indentation, and curvature. Likewise, the software subsystem can learn to map from geometric deformation to mechanical state, effectively extending the hardware's perceptual reach without adding physical sensors. The resulting system is not only more efficient but also more perceptually aligned to the human sense of touch.

5.5 Conclusion

The progression from hardware to software marks the deepening integration of the haptic system's three subsystems introduced in Chap. 1. Each subsystem—hardware, software, and application—realizes the same foundational principles through a different functional role. Hardware defines what can be sensed: the physical scope and fidelity of tactile cues. Software defines how those cues acquire meaning by way of the inference processes that translate mechanical data into perceptual states. Together, these subsystems establish the infrastructure for perception-aware interaction, forming a cohesive architecture modeled after the human somatosensory pathway. Within this framework, inference replaces stimulus replay or rule-based control as the central organizing logic, unifying sensing, interpretation, and actuation into a continuous perceptual loop.

This integration prepares the ground for the application subsystem, where perceptual inference and multimodal sensing are expressed in specific design contexts. The following chapter turns to one such context: artificial skin, which embodies the complete realization of an integrated haptic system. As an application, it extends the principles demonstrated here—afferent-level sensing, perceptual inference, and adaptive control—into a tangible interface that both measures and reasons about touch. In doing so, it translates the general architecture developed across the hardware and software subsystems into concrete engineering practice, guiding the design of haptic systems that both measure and reason about touch.

References

1. Adams, R.A., Shipp, S., Friston, K.J.: Predictions not commands: active inference in the motor system. Brain Struct. Function **218**(3), 611–643 (2013)
2. Bowden, J.L., McNulty, P.A.: Age-related changes in cutaneous sensation in the healthy human hand. Age **35**, 1077–1089 (2013)
3. Burton, H., Abend, N., MacLeod, A.M., Sinclair, R., Snyder, A., Raichle, M.: Tactile attention tasks enhance activation in somatosensory regions of parietal cortex: a positron emission tomography study. Cerebral Cortex **9**(7), 662–674 (1999)

4. Chalk, M., Marre, O., Tkačik, G.: Toward a unified theory of efficient, predictive, and sparse coding. Proc. Natl. Acad. Sci. **115**(1), 186–191 (2018)
5. Chapman, C.E., Meftah, E.M.: Independent controls of attentional influences in primary and secondary somatosensory cortex. J. Neurophysiol. **94**(6), 4094–4107 (2005)
6. Cody, T., Adams, S., Beling, P.A.: Empirically measuring transfer distance for system design and operation. IEEE Syst. J. **16**(3), 4962–4973 (2022)
7. Corniani, G., Casal, M.A., Panzeri, S., Saal, H.P.: Population coding strategies in human tactile afferents. PLOS Comput. Biol. **18**(12), e1010763 (2022)
8. Deflorio, D., Di Luca, M., Wing, A.M.: Skin properties and afferent density in the deterioration of tactile spatial acuity with age. J. Physiol. **601**(3), 517–533 (2023)
9. Edmondson, L.R., Rodríguez, A.J., Saal, H.P.: Expansion and contraction of resource allocation in sensory bottlenecks. eLife **11**, e70777 (2022)
10. Gomez-Ramirez, M., Hysaj, K., Niebur, E.: Neural mechanisms of selective attention in the somatosensory system. J. Neurophysiol. **116**(3), 1218–1231 (2016)
11. Greenacre, M., Groenen, P.J., Hastie, T., d'Enza, A.I., Markos, A., Tuzhilina, E.: Principal component analysis. Nature Rev. Methods Primers **2**(1), 100 (2022)
12. Hauser, S.C., Gerling, G.J.: Force-rate cues reduce object deformation necessary to discriminate compliances harder than the skin. IEEE Trans. Hap. **11**(2), 232–240 (2017)
13. Juravle, G., Binsted, G., Spence, C.: Tactile suppression in goal-directed movement. Psychonomic Bull. Rev. **24**(4), 1060–1076 (2017)
14. Li, B., Gerling, G.J.: Individual differences impacting skin deformation and tactile discrimination with compliant elastic surfaces. In: 2021 IEEE World Haptics Conference (WHC), pp. 721–726. IEEE (2021)
15. Li, B., Gerling, G.J.: An individual's skin stiffness predicts their tactile discrimination of compliance. J. Physiol. **601**(24), 5777–5794 (2023)
16. Li, B., Hauser, S., Gerling, G.J.: Identifying 3-d spatiotemporal skin deformation cues evoked in interacting with compliant elastic surfaces. In: 2020 IEEE Haptics Symposium (HAPTICS), pp. 35–40. IEEE (2020)
17. Li, B., Hauser, S.C., Gerling, G.J.: Faster indentation influences skin deformation to reduce tactile discriminability of compliant objects. IEEE Trans. Hap. **16**(2), 215–227 (2023)
18. Li, B., Gerling, G.J., Cody, T.: Estimating contact force rate using skin deformation cues. In: International Conference on Human Haptic Sensing and Touch Enabled Computer Applications, pp. 464–476. Springer (2024)
19. Lorena, A.C., Garcia, L.P., Lehmann, J., Souto, M.C., Ho, T.K.: How complex is your classification problem? A survey on measuring classification complexity. ACM Comput. Surv. **52**(5), 1–34 (2019)
20. Manita, S., Suzuki, T., Homma, C., Matsumoto, T., Odagawa, M., Yamada, K., Ota, K., Matsubara, C., Inutsuka, A., Sato, M., et al.: A top-down cortical circuit for accurate sensory perception. Neuron **86**(5), 1304–1316 (2015)
21. Olshausen, B.A., Field, D.J.: Sparse coding of sensory inputs. Curr. Opinion Neurobiol. **14**(4), 481–487 (2004)
22. Omrani, M., Murnaghan, C.D., Pruszynski, J.A., Scott, S.H.: Distributed task-specific processing of somatosensory feedback for voluntary motor control. eLife **5**, e13141 (2016)
23. Ostry, D.J., Darainy, M., Mattar, A.A., Wong, J., Gribble, P.L.: Somatosensory plasticity and motor learning. J. Neurosci. **30**(15), 5384–5393 (2010)
24. Parr, T., Friston, K.J.: The anatomy of inference: generative models and brain structure. Front. Comput. Neurosci. **12**, 90 (2018)
25. Parr, T., Pezzulo, G., Friston, K.J.: Active Inference: The Free Energy Principle in Mind, Brain, and Behavior. MIT Press, Cambridge (2022)
26. Pezzulo, G., Parr, T., Friston, K.: Active inference as a theory of sentient behavior. Biol. Psychol. **186**, 108741 (2024)
27. Steinley, D.: K-means clustering: a half-century synthesis. British J. Math. Stat. Psychol. **59**(1), 1–34 (2006)

28. Wark, B., Lundstrom, B.N., Fairhall, A.: Sensory adaptation. Curr. Opinion Neurobiol. **17**(4), 423–429 (2007)
29. Xia, R., Chen, X., Engel, T.A., Moore, T.: Common and distinct neural mechanisms of attention. Trends Cogn. Sci. **28**(6), 554–567 (2024)
30. Yu, Y., Huber, L., Yang, J., Jangraw, D.C., Handwerker, D.A., Molfese, P.J., Chen, G., Ejima, Y., Wu, J., Bandettini, P.A.: Layer-specific activation of sensory input and predictive feedback in the human primary somatosensory cortex. Sci. Adv. **5**(5), eaav9053 (2019)

Chapter 6
Artificial Skin as an Integrated Haptic System

Abstract Artificial skin represents the culmination of inference-centric haptic system design: an engineered embodiment of how hardware, software, and application subsystems cooperate to reproduce the perceptual and functional logic of natural touch. This chapter positions artificial skin as both a technological construct and a systems-engineering case study, illustrating how the integration of sensing, inference, and control transforms tactile surfaces into intelligent perceptual interfaces. Through the focused example of compliance perception, the chapter demonstrates how a high-level system goal can be decomposed into coordinated subsystem requirements grounded in first principles of haptic science. It further generalizes this framework across application contexts—from task-specific inspection to social and healthcare robotics—showing how the same inference-centric pipeline can scale in complexity, adapt to multimodal cognition, and ultimately support embodied intelligence.

6.1 Bringing the Application into Design

Artificial skin is an engineered sheet of compliant material designed to reproduce the essential mechanical, protective, and sensory functions of natural skin tissue [21, 22, 36]. Unlike conventional flexible sensor arrays, artificial skin is distinguished by its ability to conform to complex, arbitrary surfaces and to achieve high-resolution transduction approaching the functional envelope of biological skin. This capability holds transformative potential across diverse domains. In machine health monitoring, conformable skin layers can map strain, pressure, and temperature across irregular geometries, detecting structural anomalies before they evolve into failures [15, 41]. In teleoperation and remote communication, tactile skins establish bidirectional feedback channels that allow operators not only to command distant manipulators but also to feel remote forces, textures, and contact events in real time [5, 27]. In virtual and augmented environments, they enable immersive, full-surface haptic feedback, while in robotics, they provide machines with proprioceptive awareness, dexterous manipulation, and socially appropriate touch.

B. Li, T. Cody, *Design Principles for Integrated Haptic Systems*, Springer Series on Touch and Haptic Systems, https://doi.org/10.1007/978-3-032-16986-0_6

Artificial skin also embodies the integrated haptic system design process developed throughout this book. It exemplifies how the hardware, software, and application subsystems can be integrated together to replicate the perceptual and functional capabilities of natural touch. The hardware subsystem captures the raw physical and mechanical variables underlying tactile interaction—forces, deformations, vibrations, and other multimodal signals that encode the structure of contact events. The software subsystem receives this information and performs computational inference, transforming the raw data into perceptually meaningful representations. The application subsystem then employs these perceptual representations to fulfill the overall design goals of the haptic system. It implements control strategies, closed-loop feedback, and system-level coordination, ensuring that tactile information is used effectively to regulate behavior, adapt to changing conditions, and produce responses consistent with human-like perceptual logic. In this structure, hardware provides data, software provides interpretation, and the application subsystem provides action—together forming an iterative cycle of inference that enables the system to sense, interpret, and respond.

In this case study, compliance perception serves as a focused design goal through which this integration is examined. From a system-level perspective, compliance perception defines a target haptic capability for the artificial skin: the ability to sense, interpret, and respond to material compliance in a manner consistent with human tactile perception. Achieving this goal requires decomposing the system goal into functional requirements for each subsystem. The hardware must sense informative skin-deformation cues [17, 19, 20]; the software must extract and model those cues into perceptually relevant representations; and the application subsystem must use these representations to drive control behaviors that align with perceived compliance.

This chapter traces how these subsystems realize the design goal of compliance perception. By linking physics to perception, and perception to control, the artificial skin system illustrates how the integrated haptic system design process can be used to create inference-centric haptic systems that transform raw physical signals into structured, meaningful, and actionable information.

First, the chapter demonstrates how compliance perception can be implemented as a design goal for the integrated haptic system design process, serving to satisfy the higher-level system goal-beginning with the system-level goal and proceeding through the decomposition of subsystem requirements derived from first principles of haptic science. Then, the chapter explores how the same framework generalizes across application contexts such as human-robot interaction and healthcare, where the system-level goals differ but the design process remains consistent. The chapter concludes with a discussion of classes of machine learning methodologies that are especially relevant to inference-centric haptic systems.

Through this case study, this chapter emphasizes that adopting a systems-engineering mindset—where each design decision is guided by the flow of inference across subsystems—enables haptic technology to evolve beyond isolated input-output components toward a truly integrated, perception-aligned system. In such a system, changing the high-level goal naturally redefines the structure and requirements of its subsystems, demonstrating how integrated haptic system design can scale from tactile perception to full embodied intelligence.

6.2 Integrated Haptic System Design Process for Artificial Skin

To realize sophisticated tactile capabilities in engineered systems, haptic functionality must be developed with a clear understanding of how it contributes to broader system performance. In this section, we walk through the four-step, integrated haptic system design process for building an inference-centric haptic system, using compliance perception in artificial skin as a running example. This approach demonstrates how system-level goals can be translated into coordinated hardware, software, and application subsystems that work together to deliver perceptually grounded, functionally meaningful touch-based interactions.

6.2.1 Step 1: Identify the Haptic System Design Goals

The first step begins with the overall system-level goal—the intended performance or behavior of the overall system—and defines what specific capabilities the haptic system must enable to make that goal achievable. These desired capabilities become the haptic design goals.

In our case study, the system-level goal is: *"A robot can sort fruits in a grocery store."*

This task requires the robot to distinguish between objects based not only on vision or shape, but also on compliance—how soft each fruit is. Grapes must not be crushed, bananas must be gently gripped, and avocados must be sorted by ripeness. To support this task, the artificial skin must provide the robot with a human-like ability to sense and interpret material compliance through contact interactions, thereby enabling context-sensitive grasping and manipulation.

Thus, the design goal of the haptic system is established: *"Enable the system to perceive and discriminate object compliance in a manner consistent with human tactile perception."*

Note that other haptic goals (e.g., slip detection, texture sensing, thermal feedback) could also be layered into the system depending on the broader application. But for this example, compliance perception serves as the focused design goal through which the full integration process is explored.

6.2.2 Step 2: Identify Relevant First Principles

Once the haptic system design goal is established, the next step is to identify the first principles that govern how this capability can be achieved. These principles encompass both physical and perceptual foundations—laws and mechanisms that determine how the phenomenon can be realized and constrained.

For compliance perception, these first principles draw directly from the two scientific, foundational principles established earlier in the book:

- Foundational Principle 1: The link between skin mechanics and psychophysical responses, which defines how mechanical events on the skin is transduced into perceptual meaning. These relationships determine which physical cues must be captured and how they combine to form a percept.
- Foundational Principle 2: The effect of individual differences in tactile perception, which explains how factors such as finger size and skin stiffness can alter cue weighting and perceptual thresholds. These insights inform how models and hardware parameters should accommodate variability across users, tasks, and environments.

These first principles provide the scientific foundation for design choices, guiding what the system must measure, how data must be interpreted, and what kind of representations are needed to support perception.

6.2.3 Step 3: Decompose the Goals into Subsystems

The next step is to map the design goal—compliance perception—onto each of the three subsystems: hardware, software, and application. This decomposition ensures that all parts of the system contribute to realizing the goal in a coordinated and complementary way.

Hardware Subsystem The hardware subsystem provides the physical interface between the artificial skin and the environment. Its functional role is to capture the mechanical cues of contact events that are informative for inferring compliance. These cues include the spatial and temporal change of contact area, surface curvature, penetration depth, contact shape, and contact force—variables that collectively describe how the skin deforms under load and are correlated with compliance perception.

- To support compliance perception, the hardware must be purposefully engineered to ensure both perceptual relevance and data quality for downstream inference. First, the artificial skin must exhibit mechanical compliance that closely mimics the mechanical characteristics of biological skin—including elasticity, large-area conformability, and protective buffering. This ensures that the skin can deform naturally under load (within the perceptual range necessary for compliance discrimination—typically between 10 and 500 kPa.), conform to curved or irregular surfaces, and withstand repeated contact without compromising performance. Second, the system must ensure sensor fidelity, meaning that sensing units are capable of resolving both static and dynamic deformations with high spatial resolution (on the order of millimeters) and fast temporal response (in milliseconds), allowing the capture of transient contact events as well as steady-state behaviors. Third, the hardware must support multimodal sensing, with the ability to detect multi-axial forces, a wide spectrum of vibrations, and other modalities such as temperature or wetness [2, 34, 40]. This enables the system to distinguish between different modes of interaction, such as pure indentation versus sliding or squeezing. Finally, the system must provide signal stability and repeatability—producing consistent output across repeated contacts under similar conditions—to enable robust model training and perceptual inference. Together, these design goals define the physical sensing foundation upon which perceptually grounded, inference-centric compliance estimation can be built.

- In the fruit-sorting example, these hardware design goals translate into an artificial fingertip or palm skin that can deform around each fruit, sense distributed pressure and skin deformation, and provide quality data streams for the software subsystem to interpret.

Software Subsystem The software subsystem interprets the sensor data provided by the hardware and transforms it into perceptually meaningful representations. It serves as the computational core of the haptic system, performing inference that links measurable deformation cues to higher-level perceptual constructs such as compliance, softness, or slip. Beyond direct perception modeling, the software can also predict unmeasurable or latent cues—for instance, estimating contact force or internal strain fields from observed deformation patterns. In doing so, it bridges the gap between raw physical signals and perceptual understanding, enabling the system to reason about touch in ways that align with human sensory interpretation.

- To support compliance perception, the software subsystem must perform a series of critical functions that transform raw sensor data into actionable perceptual estimates. First, it carries out feature extraction, identifying and quantifying the most relevant deformation cues for compliance, such as contact area, penetration depth, and surface curvature. These features are then processed through cue integration and weighting, where multiple signals are combined using data-driven or physics-informed models that reflect how humans integrate tactile information. Importantly, these models should be capable of adaptive cue weighting, adjusting their sensitivity depending on context—for example, emphasizing different cues when interacting with softer versus stiffer materials. Next, the software performs predictive modeling, mapping the extracted features to estimates of compliance using statistical modeling or machine learning algorithms. To maintain robustness across variable conditions, the system must also support adaptability, allowing for model recalibration in response to changes in user group or environmental factors. Finally, the software subsystem must interface effectively with the application subsystem, producing interpretable outputs that can be directly used for control, decision-making, or haptic feedback rendering.
- In the fruit-sorting robot, the software subsystem transforms raw deformation data from artificial skin into interpretable compliance-related estimates that can further inform how the robot should adjust its grip or categorize each fruit.

Application Subsystem The application subsystem transforms perceptual inferences into actionable control behaviors, closing the loop between sensing, interpretation, and response through control strategies, adaptive feedback, decision-making, and system-level coordination. Whether adjusting grip force based on predicted compliance, triggering a sorting behavior, or modifying interaction parameters in real-time, the application subsystem ensures that inference-driven outputs are translated into purposeful actions that align with the system's functional objectives.

- To effectively realize compliance perception, the application subsystem must maintain contextual awareness, adjusting its behavior based on the characteristics of the task—for example, adapting its grasp depending on the fruit type, size, or position in the handling sequence. To ensure safety and resilience, the control architecture should incorporate mechanisms such as threshold-based constraints or adaptive damping structure that prevent excessive compression or unintended slippage, maintaining stable manipulation under variable conditions. In addition, the system should align tactile data with other sensory modalities, such as vision or proprioception, creating a multimodal control framework that mirrors the integrated nature of human perception and action. Finally, the subsystem must support performance monitoring, continuously evaluating control outcomes and using observed errors to refine both perceptual models and actuation strategies over time. Through these functions, the application subsystem transforms perceptual understanding into purposeful, context-sensitive behavior, allowing the artificial skin system to act intelligently and reliably within its operational environment.

6.2.4 Step 4: Iterative Refinement

Finally, the integrated system must be built, tested, and refined through iterative development cycles. While each subsystem has its own design requirements, their interdependence defines the overall intelligence of the artificial skin system. Hardware determines what can be sensed; software determines what can be inferred; and the application subsystem determines what can be done with that inference. The design process therefore proceeds through iterative feedback loops, where performance at the systems level drives adjustments at the subsystems level.

The implementation and refinement of the integrated haptic system proceed through an iterative cycle that links modeling, prototyping, and evaluation. The process begins by modeling the deformation behavior of the artificial skin and contact mechanics, establishing a physical and computational foundation for how the system interacts with compliant objects. Next, hardware prototypes are developed, including optimized sensor layouts and material layer configurations, followed by validation of their sensitivity to key deformation cues such as area expansion and curvature change. In parallel, the software inference models are trained and validated using psychophysical data, ensuring that the system's perceptual predictions align with human compliance perception. Finally, the application subsystem is evaluated under real-world task conditions—for instance, assessing sorting accuracy or grasp success rate—to determine how effectively the system translates perceptual inference into functional behavior.

These iterations aim to minimize inference error—the discrepancy between the system's estimated compliance and the ground-truth compliance, whether defined through physical measurement or human perception. Achieving this often requires iterative refinement across all subsystems: tuning sensor resolution and placement to enhance cue quality, adjusting inference models to improve perceptual accuracy, or revising control thresholds to optimize task performance. Through this process, the system is incrementally refined until it achieves both perceptual fidelity—that is, does the artificial skin respond in a way that mimics human tactile perception—and functional sufficiency—can the robot reliably sort fruits using compliance perception alone? In doing so, the haptic design goal becomes measurable, guiding subsystem coordination and validating whether the integrated system is capable of meeting its intended role within the larger application context.

6.3 Extending the Design to Different System-Level Goals

The integrate haptic system design process outlined in the first half of this chapter, rooted in first principles and structured through four sequential steps, is not confined to a single application or domain. Rather, it represents a generalizable framework for building integrated haptic systems across diverse contexts. The same design pipeline—from defining a system-level design goal, to identifying haptic design goals, to decomposing requirements across subsystems, and finally iterating toward performance—can be adapted to systems of varying complexity, function, and operational environment.

Now, we apply the four-step design process for a series of new system-level goals, demonstrating how the inference-centric concept translates into different real-world scenarios. Whereas the earlier case study focused on a robot sorting fruits using compliance perception, the following examples illustrate how altering the overarching goal changes every subsequent design decision—from what the haptic system must perceive, to how its hardware, software, and application subsystems are configured to achieve it.

This section aims to demonstrate the generality and flexibility of the proposed inference-centric design framework. First, it emphasizes that the system-level goal not fixed, but instead varies depending on the designer's intent, user needs, and application context. Each context defines the role of touch within the system and thus determines what the haptic subsystem must contribute. Second, when the system goal changes, so too must the subsequent steps of the design process: the definition of haptic design goals, the identification of relevant first principles, and the decomposition of subsystem requirements. An artificial skin designed to infer compliance for fruit sorting will differ markedly from one designed to detect physiological states in a healthcare setting. Third, the complexity of the haptic system can vary widely depending on the application. Some implementations may be relatively simple, involving only passive sensing or a simple prediction model (e.g., an inspection gripper), while others (such as humanoid robots) require highly integrated, adaptive systems capable of emotional prediction, bidirectional touch communication, and multimodal sensory fusion.

This section highlights the scalability and applicability of the proposed design framework. Regardless of whether the goal is to sort fruit or provide safe human-robot interaction, the same integrated haptic system design process applies: hardware captures the physical evidence of contact, software interprets that evidence into perceptual meaning, and the application subsystem transforms that meaning into context-appropriate action. By redefining the system-level goal, designers reshape the flow of inference itself, illustrating how purpose, not hardware configuration, ultimately drives the functional architecture of an integrated haptic system.

The integrated haptic system design process is intentionally general. It applies not only to haptic systems designed for controlled laboratory conditions but also to real-world contexts where the system-level goal—and thus the role of haptics—can vary dramatically. To illustrate this adaptability, two contrasting examples are presented:

- A task-specific inspection gripper
- A general-purpose humanoid robot operating in eldercare settings

These two cases differ in scope and complexity yet share the same design process: when the system goal changes, every subsequent stage of the design process—identifying first principles, defining haptic design goals, and decomposing subsystem requirements—must be revisited to ensure maintained coherence between sensing, inference, and action.

6.3.1 Scenario 1: Inspection Gripper with Contact Detection

Design Goal *A robotic gripper inspects the presence of an object.* This scenario represents a task-specific and constrained application, where the haptic system's role is specific and scoped. The robot's primary objective is to verify whether contact with an object has occurred during a vertical inspection cycle (a fixed z-axis motion with predefined speed and displacement). Unlike fruit sorting or social interaction, there is no need to perceive softness, texture, or emotional context. In this case, the haptic design goal is not to perceive nuanced material properties or textures but simply to detect contact and render normal force reliably.

First Principles The relevant first principles are those governing the mechanical relationship between indentation displacement, reaction force, and material properties. Unlike more perceptually rich tasks, psychophysical modeling is minimal; the key consideration is sensitivity to small changes in force that indicate a contact event while minimizing false positives due to vibration or noise. Signal-to-noise ratio and response latency are also critical for timely detection.

Subsystem Decomposition

- The hardware subsystem must therefore provide a compliant but mechanically stable interface capable of transducing normal forces into clear, low-noise signals. It focuses on precise force sensing along the normal axis, using a compliant artificial skin layer that can deform predictably under contact. Multimodal sensing is unnecessary; a single-axis pressure sensor may suffice; low spatial resolution is acceptable; fast temporal response should be prioritized.
- The software subsystem processes force signals to identify the onset of contact, applying simple threshold detection or low-complexity regression models to infer contact state, minimizing false detections due to vibration or frictional transients.
- The application subsystem uses this binary inference (contact or no contact), to trigger control actions—stopping the actuator, recording object presence, or signaling completion of inspection.

Iterative Refinement System refinement centers on improving detection reliability and response time. Iterative tests adjust sensor thresholds, material properties, and filtering algorithms to reduce noise and ensure rapid, stable detection under different environmental conditions.

This example highlights a simple form of inference-centric integrated haptic systems, where the hardware measures, the software interprets, and the application acts—all toward a clearly defined, task-level goal.

6.3.2 Scenario 2: Humanoid Robot in Eldercare Settings

In contrast, a humanoid robot working in a healthcare environment illustrates another end of the integrated haptic system design process—a general-purpose, socially interactive system operating in unstructured human environments. It aims to provide human-like assistance, comfort, and communication. Here, the haptic system must support physical safety, emotional expression, and social connection—capabilities that require both multifunctionality and multimodality [4, 6, 26].

Design Goal *Provide safety, trust, and empathy in interactions with human users.* This system goal is abstract. The robot may assist with physical tasks (e.g., handing over objects, repositioning a blanket), respond to human touch, and express social and emotional intent through haptic feedback. Such a system demands highly context-sensitive, adaptive, and socially competent interaction capabilities. The haptic system must contribute to this goal by enabling the robot to perceive and convey tactile information in ways that feel natural and appropriate to human users. The haptic design goal extends beyond simple contact detection to encompass multi-function and multimodal tactile perception, see Fig. 6.1. The robot must be

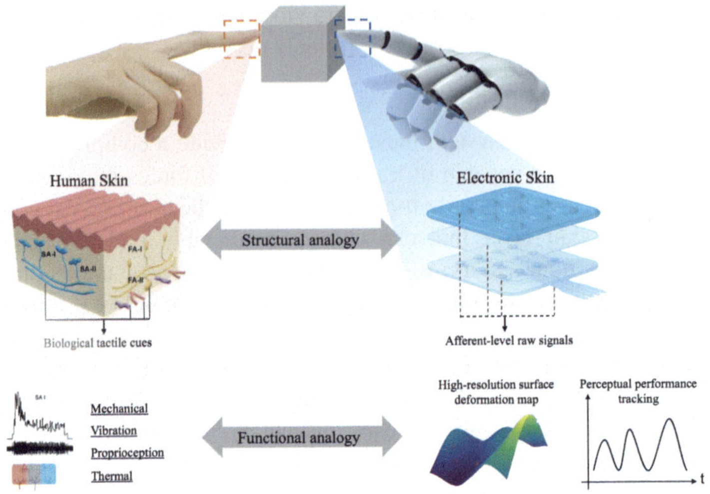

Fig. 6.1 Human skin (left) and artificial skin (right) share multifunctional, multimodal sensing capabilities, enabling detection of diverse tactile modalities such as pressure, vibration, motion, and temperature. Both systems achieve high spatial and temporal resolution, allowing fine-grained, real-time perception of complex contact events

capable of perceiving compliance, friction, temperature, and motion cues, while also recognizing affective patterns of touch—gentle pressure, sustained holding, or stroking—that signal comfort or intention.

First Principles The first principles can be drawn from human tactile perception: these two science foundations discussed earlier in Chaps. 2 and 3:

- The link between skin mechanics and psychophysical response defines how deformation, vibration, and thermal cues encode perceptual meaning-both mechanically and socially.
- The study of individual differences in tactile perception informs how variability in touch sensitivity and preference shapes human interpretation of robotic contact.

These principles define the physical range, perceptual resolution, and emotional valence the robot's artificial skin must replicate to achieve human-like expressivity and responsiveness.

Subsystem Decomposition

- The hardware subsystem requires multi-functional, multimodal sensing and actuation, capable of detecting and rendering not only normal and shear forces but also vibration, temperature, and possibly affective cues such as pressure rhythm or stroke velocity. The skin must maintain softness and compliance for safe, comfortable contact while integrating dense sensor arrays for high spatial resolution.

- The software subsystem must perform multi-level inference, transforming raw sensor data into perceptual constructs (e.g., softness, slip, warmth) and higher-order interpretations such as comfort, reassurance, or alert. The software pipeline should manage cue extraction, affective inference, and contact classification, allowing the system to interpret both the physical and emotional dimensions of touch. It must employ real-time models that can distinguish between different types of tactile interactions—such as an intentional touch like hand-holding, an accidental bump, or an urgent contact indicating distress—ensuring that each is met with an appropriate behavioral response. To further enhance personalization and trust in human–robot interaction, the software should incorporate adaptive learning mechanisms that enable user-specific calibration, accounting for variations in touch pressure, duration, or expressive style among different individuals.
- The application subsystem uses these perceptual and emotional inferences to regulate behavior: adjusting grip during assistance, synchronizing tactile feedback with verbal communication, or expressing reassurance through gentle touch. Control strategies involve both safety mechanisms (collision avoidance, pressure limits) and social modulation (context-aware gesture timing and force).

Iterative Refinement Because the system operates in highly variable social and physical environments, iteration involves long-term adaptation. Continuous learning updates perceptual models based on user feedback and contextual cues, while control parameters evolve to maintain both safety and emotional congruence. Over time, the system develops an embodied "tactile intelligence" that balances technical reliability with emotional appropriateness—achieving the higher-level goal of safe, trustworthy, and empathetic care.

This example illustrates a high-complexity, perception-rich haptic system. The artificial skin here is not only a sensing surface, but also a social and perceptual interface, enabling the robot to operate in human-centered, emotionally nuanced environments. The same design process guides the development, but the goals, constraints, and integration strategies are fundamentally different from the inspection gripper scenario.

6.3.3 Design Implications

These two examples illustrate how the integrated haptic system design process scales across different contexts. The inspection gripper demonstrates a scoped, task-specific implementation, while the humanoid robot embodies a complex, perception-driven system in which inference spans mechanical, cognitive, and affective dimensions.

In both cases, the four-step integrated haptic system design process remains the same: identify the haptic system design goals, ground them in first principles, decompose them across subsystems, and refine the design iteratively. What changes is the scope and depth of inference. In simple systems, inference terminates in binary decisions; in complex systems, it cascades through layers of perception, interpretation, and social behavior. This flexibility underscores the generality and scalability of the proposed design framework: by anchoring design to inference-related, closed-loop needs of the application subsystem rather open-loop sensing and actuation, the integrated haptic system design process can evolve haptic systems from basic sensing tools to future haptic systems that enable embodied, human-aligned intelligent systems.

6.4 Supporting System-Level Inference

Within the integrated haptic system, the application subsystem functions as the architectural interface between tactile perception and system-level cognition. It defines the representational and operational boundaries through which tactile inference contributes to broader reasoning processes. In practice, this subsystem models the key interactions between the haptic system and the cognitive architecture—what tactile representations are exposed, what control commands are accepted, and how the two are synchronized.

By formalizing these exchanges early in the design process, the application subsystem ensures that system-level needs—such as multimodal fusion, task reasoning, or safety validation—can be addressed through coordinated design across hardware and software. It is the workspace where perceptual intelligence meets behavioral intent of the broader system.

The inference capability of an integrated haptic system extends beyond perception. Once tactile signals are sensed, transformed, and interpreted within the hardware and software subsystems, they become part of a larger cognitive loop: an ongoing system-level process that integrates multiple sources of information—vision, language, proprioception, and touch—into coherent behavior.

The scope of this haptic-system contribution to the broader system-level goals depends on the complexity of the broader system itself. A robotic gripper represents a minimal example of system-level inference built primarily from tactile feedback. Its operational goal—securely manipulating an object—can be achieved through closed-loop tactile reasoning alone: estimating local contact state, predicting slip, and adjusting grip force accordingly. The entire inference process occurs within the tactile modality.

A healthcare robot, however, exemplifies a multimodal inference architecture. Here, tactile inference operates alongside other inferential systems: computer vision provides spatial and safety awareness, natural language models interpret verbal cues and emotional tone, and higher-level decision modules integrate these signals into socially and ethically appropriate actions [23, 31, 37]. In such systems, touch does not stand alone; it participates in a network of inference pipelines, each processing different modalities yet converging toward a shared understanding of the user and environment. The haptic system design process does not solve these multimodal challenges directly, but it provides perceptually valid and structured representations of touch that can be fused with vision, language, and intent inference at the systems level.

From this systems perspective, we can identify three dominant machine-learning paradigms that describe how haptic inference interacts with other inferential processes: representation learning, generative modeling, and reinforcement learning. Each provides a lens for understanding how perceptual intelligence in haptic subsystems scales upward into embodied, adaptive, and autonomous behavior in the broader system.

6.4.1 Building Shared Perception with Representation Learning

Representation learning concerns how perceptual information is expressed and organized so that it becomes meaningful to higher levels of reasoning [1]. In the haptic domain, this involves determining what form tactile perception takes when it leaves the software subsystem and enters the system-level inference framework. The representation may be low-level, consisting of raw voltage readings, strain distributions, or unprocessed time-series data from the artificial skin; or it may be high-level, consisting of derived biomechanical cues such as contact area, curvature, penetration depth, or even perceptual constructs such as softness and slipperiness.

The application subsystem is where these representational forms are defined and validated against system-level requirements. It specifies what the cognitive architecture needs to know about touch—whether as biomechanical vectors for grasp planning or psychophysical descriptors for affective interaction—and constrains the software subsystem to deliver data in that form. In this way, representation learning is not an emergent afterthought but a designed interface, shaped by the application subsystem to ensure semantic compatibility between haptic and non-haptic inference modules.

The more perceptually meaningful and structured these representations are, the easier it becomes for other subsystems—vision, language, control—to integrate and utilize them [12]. For instance, a robot trained on high dimensional but interpretable tactile cues can learn embeddings that relate the geometry of contact with visual features of the same object, or linguistic descriptors that characterize its texture. This process of representation alignment creates a shared perceptual space where data from multiple modalities can be compared, fused, and jointly reasoned about [18].

At the systems level, this is precisely what representation learning aims to achieve: to discover internal representations that capture the latent structure of the environment in a way that supports multiple tasks and sensory inputs. When extended to the haptic modality, representation learning enables models to generalize tactile knowledge to non-tactile contexts—recognizing materials visually, predicting compliance from sound, or describing sensations verbally. Conversely, it allows systems to infer tactile states even in the absence of touch, drawing upon learned correspondences with visual or linguistic representations.

Representation learning also provides a formal mechanism for personalization and generalization [32]. In simple systems like the gripper, the learned representation may map deformation cues directly to mechanical affordances—grasp stability, friction coefficient, or slippage probability. In complex systems like the healthcare robot, the representation may instead encode psychophysical constructs—comfort, safety, or empathy—derived from both tactile and linguistic feedback. In either case, the learned representation becomes a bridge between perception and cognition, allowing the system to reason abstractly about the meaning of touch.

Finally, the choice of representation influences not only perception but communication between subsystems. A well-structured tactile embedding acts as a common language for multimodal intelligence: a space in which vision, language, and haptics can exchange information efficiently. The success of system-level inference thus depends on the expressiveness of tactile representations and their compatibility with other sensory modalities. Representation learning, in this sense, is not just an optimization technique but a means of learning the grammar of embodied perception [8].

6.4.2 Generative Touch and Tactile Prompting

While representation learning focuses on how sensory information is encoded, generative modeling concerns what sensations, actions, or contextual expectations a given touch should elicit [11, 30]. In haptic systems, generative modeling introduces the idea that tactile events can act as prompts—inputs that elicit multimodal predictions, behaviors, or feedback [9, 25, 38]. A tactile interaction, such as a gentle touch on the arm, is not a static observation but a dynamic signal that carries implications about context, intent, and meaning.

In the robotic gripper, generative modeling operates locally. Before physical contact, the system may simulate probable deformation patterns or slip trajectories given the object's shape and material, thereby generating tactile expectations that guide control. These expectations reduce uncertainty and improve efficiency—much as human motor systems predict proprioceptive feedback before movement. The generative process here is narrow but complete: it predicts the next tactile state, compares it against actual feedback, and corrects the model in real time.

In contrast, the healthcare robot operates in a deeply multimodal generative context [7, 10, 33, 35]. Here, tactile events are coupled with visual and linguistic information: a patient's gesture, facial expression, or verbal feedback all co-occur with touch. The system can use tactile data to generate hypotheses about emotional state ("the patient is tense"), safety ("the contact pressure is excessive"), or social meaning ("the user is initiating comfort touch"). Conversely, it can generate tactile responses—adjusting pressure, warmth, or duration—based on linguistic or visual cues.

Inference-centric haptic systems support this bidirectionality. Tactile inputs can serve as conditioning variables: the tactile state provides a "prompt" that constrains how other modalities are generated or interpreted. For instance, a healthcare robot might use tactile prompts to generate verbal explanations ("Does this pressure feel comfortable?") or to guide visual attention ("Focus on the region being massaged"). Furthermore, generative modeling enables the system to perform touch synthesis: creating artificial tactile feedback consistent with multimodal context. Given a description, an image, or a recorded interaction, the model can generate a corresponding tactile signal to replay through actuators. Such generative capabilities allow haptic systems to move from reactive to creative—capable of producing new tactile experiences that are coherent with environmental and communicative goals.

6.4.3 Reinforcement Learning and Touch

The third major paradigm through which system-level inference engages with haptics is reinforcement learning (RL) [28], which focuses on learning through interaction with the environment. Reinforcement learning provides a natural framework for optimizing tactile behavior, because touch is inherently interactive: it changes the environment even as it perceives it [3, 29].

The application subsystem plays a critical role in defining the state and reward representations that connect tactile experience to system-level learning. It contributes tactile variables to be included in the policy space, inputs that shape rewards, and determines how updates to the psychophysical inference model are triggered by reinforcement learning outcomes. Through this interface, the subsystem enables two-way adaptation: system-level policies refine tactile interpretation, while tactile experiences inform system-level learning.

In simple systems like the robotic gripper, reinforcement learning operates within the tactile domain itself. The RL agent receives tactile feedback as part of its state, evaluates the resulting reward (e.g., a stable grasp or successful lift), and adjusts its policy to improve future performance [16, 39]. Over time, it learns the mapping between tactile signals and effective actions without requiring explicit supervision. This creates a form of procedural inference: the system learns what to do based on what it feels.

In more complex systems, reinforcement learning operates across modalities and over extended timescales. The healthcare robot, for example, can use reinforcement learning to adapt its tactile interaction style to each individual user [13, 14, 24]. By observing both tactile feedback (pressure, temperature, compliance) and verbal or visual responses (facial expressions, verbal feedback), the system can infer the user's comfort level and adjust its behavior accordingly. Over time, the robot builds a personalized model of the user's psychophysical profile, linking perceptual cues with affective states and preferences.

This process effectively merges reinforcement learning and psychophysical inference: the system not only learns how to act to maximize reward but also refines its internal model of how tactile stimuli are perceived by that specific user. The tactile inference model thus becomes adaptive, co-evolving with the system's behavioral policy. Such continuous adaptation is essential for long-term autonomy, where interactions cannot be predefined but must be learned from experience.

At a broader level, reinforcement learning serves as the integrating mechanism among all inference processes. Representation learning defines the perceptual space in which states are recognized; generative modeling predicts outcomes and synthesizes responses; and reinforcement learning closes the loop by evaluating those predictions against real experience. Together, these mechanisms create a systems that refine both their perception and action through embodied interaction.

6.4.4 Haptic Inference in the Architecture of Intelligence

Through the lenses of representation learning, generative modeling, and reinforcement learning, we can now view the integrated haptic system as a contributor to the full cognitive architecture of intelligent systems. Each paradigm reveals a different aspect of how tactile inference supports system-level reasoning: representation learning builds common perceptual structure, generative modeling extends prediction and communication, and reinforcement learning optimizes adaptation through experience.

In simple systems, such as the gripper, these mechanisms operate locally and autonomously: perception leads directly to action. In complex systems, such as the healthcare robot, they interlock across modalities and cognitive layers, linking physical touch with visual perception, natural language understanding, and social reasoning. The haptic system provides the grounding that keeps these higher-level processes connected to physical reality, ensuring that learning and inference remain embodied rather than abstract. Ultimately, the application subsystem serves as the translation layer that binds perceptual inference to cognitive inference. It is both an engineering artifact and a conceptual bridge, ensuring that haptic perception is not a closed circuit but a contributor to system-level understanding. Viewed from this perspective, the artificial skin is more than a sensing surface—it is an interface between perception and cognition. It provides structured, perceptually aligned information that higher-level systems can learn from, reason with, and act upon. As integrated haptic systems continue to evolve, their contribution to system-level inference will deepen, enabling systems not just to simulate touch but to use it as a means of knowing and adapting.

6.5 Conclusion

The artificial-skin case study demonstrates that the unification of hardware, software, and application subsystems produces more than a sensitive surface—it establishes a platform for embodied inference. Each subsystem contributes a distinct layer of intelligence: hardware captures the mechanical evidence of touch, software infers its perceptual meaning, and the application subsystem coordinates these in service of system-level goals. Together, they form a continuous perceptual loop that aligns engineering structure with the logic of natural sensation. At the systems level, the artificial skin becomes a bridge between physical interaction and cognitive reasoning. Through its structured representations, predictive capabilities, and adaptive control, it enables machines not only to detect contact but to understand and participate in it. In doing so, artificial skin transforms haptics from a peripheral modality into a central participant in embodied cognition—illustrating how the design of touch can inform the design of intelligence itself.

As the examples in this chapter show, redefining the system-level goal reconfigures this loop. A simple inspection gripper compresses inference into binary contact reasoning, whereas a healthcare robot extends it into multimodal and social cognition. Yet both adhere to the same inference-centric design process: sense, interpret, and act through integrated subsystems. This scalability is what makes the integrated haptic system design process general—it does not prescribe a single form of touch but a way of organizing intelligence around it.

References

1. Bengio, Y., Courville, A., Vincent, P.: Representation learning: A review and new perspectives. IEEE Trans. Pattern Anal. Mach. Intell. **35**(8), 1798–1828 (2013)
2. Cao, V.A., Nguyen, N.K., Kim, M., Van, P.C., Minh, H.N., Kim, S.Y., Nah, J., et al.: Multifunctional tactile sensor with multimodal capabilities for pressure, temperature, and surface recognition. Nano Energy **136**, 110706 (2025)
3. Church, A., Lloyd, J., Hadsell, R., Lepora, N.F.: Deep reinforcement learning for tactile robotics: learning to type on a braille keyboard. IEEE Robot. Autom. Lett. **5**(4), 6145–6152 (2020)
4. Cucciniello, I., Sangiovanni, S., Maggi, G., Rossi, S.: Mind perception in hri: Exploring users' attribution of mental and emotional states to robots with different behavioural styles. Int. J. Soc. Robot. **15**(5), 867–877 (2023)
5. Dangxiao, W., Yuan, G., Shiyi, L., Weiliang, X., Jing, X., et al.: Haptic display for virtual reality: progress and challenges. Virtual Reality Intell. Hardware **1**(2), 136–162 (2019)
6. Davis, J.L., Armstrong, R., Groggel, A., Doolan, S., Sheedy, J., Love, T.P., Herath, D.: Role-taking and robotic form: an exploratory study of social connection in human-robot interaction. Int. J. Human-Comput. Stud. **178**, 103094 (2023)
7. Dou, Y., Yang, F., Liu, Y., Loquercio, A., Owens, A.: Tactile-augmented radiance fields. In: Proceedings of the IEEE/CVF Conference on Computer Vision and Pattern Recognition, pp. 26529–26539 (2024)
8. Du, Y., Gan, C., Isola, P.: Curious representation learning for embodied intelligence. In: Proceedings of the IEEE/CVF International Conference on Computer Vision, pp. 10408–10417 (2021)
9. Faruqi, F., Perroni-Scharf, M., Walia, J.S., Zhu, Y., Feng, S., Degraen, D., Mueller, S.: Tactstyle: Generating tactile textures with generative ai for digital fabrication. In: Proceedings of the 2025 CHI Conference on Human Factors in Computing Systems, pp. 1–16 (2025)

10. Gao, R., Yuan, W., Zhu, J.Y.: Controllable visual-tactile synthesis. In: Proceedings of the IEEE/CVF International Conference on Computer Vision, pp. 7040–7052 (2023)
11. Goodfellow, I., Pouget-Abadie, J., Mirza, M., Xu, B., Warde-Farley, D., Ozair, S., Courville, A., Bengio, Y.: Generative adversarial networks. Commun. ACM **63**(11), 139–144 (2020)
12. Guo, W., Wang, J., Wang, S.: Deep multimodal representation learning: a survey. IEEE Access **7**, 63373–63394 (2019)
13. Guzey, I., Dai, Y., Evans, B., Chintala, S., Pinto, L.: See to touch: Learning tactile dexterity through visual incentives. In: 2024 IEEE International Conference on Robotics and Automation (ICRA), pp. 13825–13832. IEEE (2024)
14. Hansen, J., Hogan, F., Rivkin, D., Meger, D., Jenkin, M., Dudek, G.: Visuotactile-rl: Learning multimodal manipulation policies with deep reinforcement learning. In: 2022 International Conference on Robotics and Automation (ICRA), pp. 8298–8304. IEEE (2022)
15. Hassan, M.F., Li, Z., Islam, M.S., Gencturk, B., Zheng, B., Pan, X., Khan, Y., Muin, S.: Wireless printed large-area sensors for continuous structural health monitoring. Adv. Mater. Technol. **10**(5), 2401782 (2025)
16. Hellman, R.B., Tekin, C., Van Der Schaar, M., Santos, V.J.: Functional contour-following via haptic perception and reinforcement learning. IEEE Trans. Hap. **11**(1), 61–72 (2017)
17. Li, B., Gerling, G.J.: Individual differences impacting skin deformation and tactile discrimination with compliant elastic surfaces. In: 2021 IEEE World Haptics Conference (WHC), pp. 721–726. IEEE (2021)
18. Li, Y., Yang, M., Zhang, Z.: A survey of multi-view representation learning. IEEE Trans. Knowl. Data Eng. **31**(10), 1863–1883 (2018)
19. Li, B., Hauser, S., Gerling, G.J.: Identifying 3-d spatiotemporal skin deformation cues evoked in interacting with compliant elastic surfaces. In: 2020 IEEE Haptics Symposium (HAPTICS), pp. 35–40. IEEE (2020)
20. Li, B., Lee, J., Gerling, G.J.: A magnetorheological elastomer device for programmable actuation and sensing of soft haptic experiences. In: 2021 IEEE World Haptics Conference (WHC), pp. 349–349. IEEE (2021)
21. Liu, Z., Hu, X., Bo, R., Yang, Y., Cheng, X., Pang, W., Liu, Q., Wang, Y., Wang, S., Xu, S., et al.: A three-dimensionally architected electronic skin mimicking human mechanosensation. Science **384**(6699), 987–994 (2024)
22. Lucarotti, C., Oddo, C.M., Vitiello, N., Carrozza, M.C.: Synthetic and bio-artificial tactile sensing: a review. Sensors **13**(2), 1435–1466 (2013)
23. Mazursky, A., DeVoe, M., Sebo, S.: Physical touch from a robot caregiver: Examining factors that shape patient experience. In: 2022 31st IEEE International Conference on Robot and Human Interactive Communication (RO-MAN), pp. 1578–1585. IEEE (2022)
24. Pecyna, L., Dong, S., Luo, S.: Visual-tactile multimodality for following deformable linear objects using reinforcement learning. In: 2022 IEEE/RSJ International Conference on Intelligent Robots and Systems (IROS), pp. 3987–3994. IEEE (2022)
25. Soh, H., Demiris, Y.: Incrementally learning objects by touch: Online discriminative and generative models for tactile-based recognition. IEEE Trans. Hap. **7**(4), 512–525 (2014)
26. Stock-Homburg, R.: Survey of emotions in human–robot interactions: Perspectives from robotic psychology on 20 years of research. Int. J. Soc. Robot. **14**(2), 389–411 (2022)
27. Stone, R.J.: Haptic feedback: a brief history from telepresence to virtual reality. In: International Workshop on Haptic Human-Computer Interaction, pp. 1–16. Springer (2000)
28. Sutton, R.S., Barto, A.G., et al.: Reinforcement Learning: An Introduction, vol. 1. MIT Press Cambridge (1998)
29. Van Hoof, H., Chen, N., Karl, M., Van Der Smagt, P., Peters, J.: Stable reinforcement learning with autoencoders for tactile and visual data. In: 2016 IEEE/RSJ International Conference on Intelligent Robots and Systems (IROS), pp. 3928–3934. IEEE (2016)
30. Vaswani, A., Shazeer, N., Parmar, N., Uszkoreit, J., Jones, L., Gomez, A.N., Kaiser, Ł., Polosukhin, I.: Attention is all you need. Adv. Neural Inf. Process. Syst. **30** (2017)
31. Wang, T., Zheng, P., Li, S., Wang, L.: Multimodal human–robot interaction for human-centric smart manufacturing: a survey. Adv. Intell. Syst. **6**(3), 2300359 (2024)

32. Wang, R., Zhao, D., Suh, D., Yuan, Z., Chen, G., Min, B.C.: Personalization in human-robot interaction through preference-based action representation learning. In: 2025 IEEE International Conference on Robotics and Automation (ICRA), pp. 7377–7384. IEEE (2025)
33. Xi, Q., Wang, F., Tao, L., Zhang, H., Jiang, X., Wu, J.: Cm-avae: Cross-modal adversarial variational autoencoder for visual-to-tactile data generation. IEEE Robot. Autom. Lett. **9**(6), 5214–5221 (2024)
34. Xie, Y., Pan, J., Yu, L., Fang, H., Yu, S., Zhou, N., Tong, L., Zhang, L.: Optical micro/nanofiber enabled multiaxial force sensor for tactile visualization and human–machine interface. Advanced Science **11**(45), 2404343 (2024)
35. Zhang, T., Cong, Y., Sun, G., Dong, J., Liu, Y., Ding, Z.: Generative partial visual-tactile fused object clustering. In: Proceedings of the AAAI Conference on Artificial Intelligence, vol. 35, pp. 6156–6164 (2021)
36. Zhang, M., Xing, J., Zhong, Y., Zhang, T., Liu, X., Xing, D.: Advanced function, design and application of skin substitutes for skin regeneration. Mater. Today Bio **24**, 100918 (2024)
37. Zhao, W., Gangaraju, K., Yuan, F.: Multimodal perception-driven decision-making for human-robot interaction: a survey. Front. Robot. AI **12**, 1604472 (2025)
38. Zhong, S., Albini, A., Maiolino, P., Posner, I.: Tactgen: tactile sensory data generation via zero-shot sim-to-real transfer. IEEE Trans. Robot. **41**, 1316–1328 (2024)
39. Zhou, Y., Jin, Y., Lu, P., Jiang, S., Wang, Z., He, B.: T-td3: A reinforcement learning framework for stable grasping of deformable objects using tactile prior. IEEE Trans. Autom. Sci. Eng. **22**, 6208–6222 (2024)
40. Zhu, H., Luo, H., Cai, M., Song, J.: A multifunctional flexible tactile sensor based on resistive effect for simultaneous sensing of pressure and temperature. Adv. Sci. **11**(6), 2307693 (2024)
41. Zymelka, D., Togashi, K., Kobayashi, T.: Carbon-based printed strain sensor array for remote and automated structural health monitoring. Smart Mater. Struct. **29**(10), 105022 (2020)

Chapter 7
The Future of Integrative Haptics

Abstract This chapter advances a unified philosophy of integrative haptics, proposing that the future of haptic design lies in achieving informational coherence across sensing, inference, and application. It identifies the haptic information bottleneck—the asymmetry between biological and artificial tactile bandwidths–as the primary constraint on perceptual realism, and introduces haptic accessibility as the condition in which this gap is minimized through parity in representation and response. Extending beyond parity, haptic reciprocity emerges as the mutual adaptation of human and artificial systems within shared inferential loops, culminating in the vision of universal tactile interaction—a framework where touch functions as a common medium of cognition, coordination, and understanding.

7.1 Toward a Philosophy of Integrated Haptic Design

The culmination of this book's argument invites a broader reflection. While earlier chapters established the analytical mechanics and architectures of touch, this chapter reinterprets them as a philosophical foundation for design itself. Integrated, inference-centric haptic systems compel us to think of perception and action as co-constructed phenomena—every sensor, algorithm, and interface becomes a medium for inference. Thus, the philosophy of integrated haptic system design is neither purely technical nor purely phenomenological; it is the study of how physical, computational, and experiential domains converge to create embodied intelligence.

By synthesizing system engineering, neuroscience, and design under a single framework, termed the integrated haptic system design process, this book proposes a new discipline: integrative haptics. Its methodology is deductive—grounded in first principles—and its application enables designers to create devices that learn, adapt, and collaborate with humans. In future haptic systems, the realism of touch will no longer be measured solely in Newtons or Hertz but in the fidelity of inference—how accurately a system can infer intent, context, and meaning from contact.

B. Li, T. Cody, *Design Principles for Integrated Haptic Systems*, Springer Series on Touch and Haptic Systems, https://doi.org/10.1007/978-3-032-16986-0_7

Integrative haptics is the interdisciplinary science and engineering of touch that unifies biomechanical sensing, computational inference, and system-level application within a closed-loop architecture. It treats haptic capability as an emergent property of coordinated hardware, software, and behavioral subsystems designed to achieve perceptually coherent, adaptive, and context-aware interaction.

Integration as the Defining Paradigm This book began by asserting that touch is not an isolated sensory channel but an interactive process between mechanics, perception, and control. Chapter 1 established that haptic capabilities are closed-loop phenomena where sensing, inference, and actuation co-evolve. The integrated haptic system was defined as a coordinated triad of hardware, software, and application subsystems whose coupling delivers perceptually meaningful feedback. The book's central thesis—that integration and inference jointly define the future of haptics—therefore emerges from systems theory itself: effective haptic design is not component optimization but architectural coherence among sensing, computation, and purposeful behavior.

From First Principles to Design Principles Chapter 2 grounded this framework in the first foundational principle of biomechanical–informational coupling, linking the deformation of skin to the information the nervous system extracts. Here, the physics of soft tissue became the formal gateway to perception. Mechanical indentation, shear, and vibration were treated not as mere inputs but as structured information fields that encode perceptual cues such as compliance, roughness, friction, and shape. This articulation translated psychophysics into quantitative design parameters.

Variation as a Design Variable Chapter 3 introduced the second foundational principle: individual differences in perception. By demonstrating how variations in finger size, ridge density, skin stiffness, and hydration alter tactile discrimination, the chapter reframed variability not as experimental noise but as an intrinsic property of any population of haptic systems—biological or engineered. Human perception is not uniform; thus, haptic design must anticipate heterogeneity in its users and contexts. This insight expands the scope of design goals from achieving nominal accuracy to achieving robust perceptual equivalence across individual differences, a theme that echoes through the book's later treatment of inference and adaptation.

Hardware as the Physical Foundation of Inference Building on these scientific foundations, Chap. 4 translated tactile physics into hardware design goals. It argued that the mechanical substrate of a haptic device is not simply a transducer array but the material embodiment of perceptual inference. By prioritizing deformation-based cues, distributing multimodal sensing, and embedding calibration mechanisms, hardware becomes the first computational layer of integrated haptic systems. Further, multifunctional and multimodal architectures are necessities for encoding the physical variables that perception depends upon. Inference begins with the material itself: the structure of sensors and actuators defines the informational affordances available to subsequent subsystems.

Software as the Medium of Perception Chapter 5 extended the argument upward into the software subsystem, which performs the transformations that make raw signals perceptually intelligible. Analogous to the afferent pathways of the nervous system, software integrates, filters, and predicts. It implements adaptive cue selection, multimodal fusion, and predictive modeling of compliance and other perceptual states.

Application as System-Level Context and Purpose Chapter 6 completed the triad by situating haptics within application systems, exemplified by the design of artificial skin. This chapter operationalized the integrated haptic system design process: (1) identify haptic system design goals; (2) map first principles; (3) decompose across subsystems; and (4) iterate. The case studies illustrated how differing system design goals reshape subsystem priorities. In the gripper, tactile inference serves manipulation accuracy; in healthcare, it serves empathy and safety through multimodal understanding. The chapter concluded that haptic inference only achieves meaning when embedded in application-level cognition, where tactile, visual, and linguistic models converge to form shared perception.

From Integration to Inference Taken together, Chapters 1–6 articulate a coherent philosophical arc: touch is an inferential system, and haptics must therefore be designed as an inferential architecture. The biological analogue—sensorimotor inference in the human body—serves not as a metaphor but as a direct framing. Hardware corresponds to peripheral encoding, software to predictive modeling, and application to contextual control. The coupling of these subsystems realizes adaptive, context-aware behavior. Thus, the book's thesis extends beyond tactile technology: it positions haptics as a paradigmatic case for designing intelligent physical systems, where perception and action are unified through continual information exchange.

The chapters leading to this point reveal a unifying insight: integrative haptics is not merely an implementation strategy, but moreover a systems worldview. When hardware, software, and application subsystems are co-designed the boundary between sensing and knowing dissolves. Touch becomes computation, and computation becomes an act of understanding—an understanding that remains incomplete so long as the informational bandwidth between human and machine is constrained. This philosophical pivot establishes the basis for the next inquiry: how can such systems overcome the constraints that presently limit their informational depth and perceptual realism?

7.2 The Haptic Information Bottleneck

Building upon this philosophical foundation for a new discipline of integrative haptics, the next challenge lies in practice: if integration and inference define the ideal, what technical barriers prevent their realization? The answer begins with the problem of information flow.

Current haptic systems remain constrained by a fundamental information bottleneck—the inability to sense, transmit, and act upon rich, high-dimensional tactile data in a meaningful way that aligns with human perceptual processes. While haptics is inherently a high-bandwidth sensory modality—driven by nuanced skin deformation and temporal dynamics—most systems can only extract and relay sparse, low-resolution signals such as discrete force values and vibrational patterns. This bottleneck prevents higher-level systems from accessing the critical biomechanical and perceptual cues necessary for intelligent and adaptive interaction.

> The **haptic information bottleneck** refers to the structural limitation that prevents current haptic systems from capturing, transmitting, and interpreting the full spatiotemporal and perceptual richness of tactile interaction. It arises from asymmetries in sensing fidelity, data bandwidth, and inference capability that restrict the flow of biomechanically and psychophysically meaningful information within or between biological and artificial haptic systems.

One key challenge is the limited transmission of biomechanically relevant data. Human tactile perception depends not just on force values (kinesthetic) but on complex spatiotemporal patterns of skin deformation (cutaneous)—such as contact area change rate and curvature gradient—which convey essential information about object properties and interaction states. Yet most current haptic devices operate with discrete force values or fixed vibration patterns, overlooking these rich cues. To bridge this gap, the development of advanced haptic hardware that mimics the spatiotemporal fidelity of the human skin becomes essential. These sensors and actuators should be capable of capturing surface deformation with millimeter-scale spatial resolution, resolving dynamic interaction variables such as contact area and shear rate while maintaining high temporal fidelity to match perceptual thresholds. By mirroring the biomechanics of touch, haptic data streams can convey the physical realities of contact that underlie user-perceived sensations.

Even when high-quality tactile data is available, there is a lack of formal mechanisms to determine which signals are perceptually important under a given context, what kind of data to transmit, when, and why. This leads to insufficient bandwidth use, redundant computation, and information overload at higher control layers where only a small subset of the input may be task-relevant. Embedding psychophysical inference models in the system provides a solution. These models dynamically assess the perceptual significance of incoming signals, extract task-critical features, and re-weight cues based on context and interactive conditions to deliver compact, perceptually aligned outputs. This mirrors attentional mechanisms in the human brain, which continuously modulate signal processing to focus on task-critical information while down-weighting predictable or redundant input. This also enables a context-aware, perceptually meaningful data handoff that avoids premature abstraction while improving communication efficiency between low-level sensing (hardware subsystem) and high-level decision making (software and application subsystems).

Another contributor to the information bottleneck is the lack of closed-loop feedback in most systems with haptic capabilities. In many implementations, actuation follows open-loop rules—delivering fixed, pre-set patterns regardless of real-time sensory feedback. This approach prevents the system from adapting to changes in contact, user behavior, or interaction goals. A perception-driven closed-loop framework overcomes this limitation by constantly comparing sensed perceptual states with defined targets, calculating perceptual error, and adjusting actuator commands accordingly. Distributed sensing and actuation units can then respond with localized, adaptive outputs that reflect the physical realities of touch. This closed-loop architecture of integrating hardware and software enables adaptive and optimal control, and ensures that actuation performance matches user expectations and environmental conditions—creating a perceptual coherence that is critical for realism and task performance in applications such as teleoperation and immersive simulation.

Rather than relying on heuristic rules or empirical tuning, the next generation of haptic systems should be designed from first principles rooted in the science of skin mechanics, psychophysics, and neural encoding. By grounding both hardware and software in these foundational principles, haptic systems can move beyond simple signal generation to transmit tactile information that is meaningful, interpretable, and perceptually aligned. This framework unifies high-resolution, real-time biomechanical sensing with inference models that mirror human perceptual coding, effectively dismantling the haptic information bottleneck across each stage of the system. As a result, the flow of data—from sensors to actuators to higher-order control modules—becomes streamlined, efficient, and grounded in perceptual goals. This enables haptic subsystems to integrate seamlessly with larger architectures such as in robotics, virtual reality, and teleoperation, where tactile feedback must coordinate with visual, auditory, and proprioceptive streams. By resolving the information bottleneck and embedding haptics as a high-bandwidth, perceptually salient sensory modality, this science-based framework transforms touch from a peripheral feature into a core component of intelligent, human-aligned interaction—empowering the next generation of intelligent systems and immersive technologies.

7.3 Haptic Accessibility and Beyond

If the haptic information bottleneck defines the present limitation of artificial systems, haptic accessibility defines the next horizon. Where the former asks how far today's systems remain below human tactile performance, the latter asks how far we might go beyond it. The central premise is that accessibility emerges when biological and artificial haptic systems achieve informational parity: when the tactile signals exchanged across the interface are encoded, transmitted, and interpreted with minimal loss of perceptual meaning. In such conditions, the human and machine no longer meet across a rigid boundary but participate in a shared sensorimotor continuum.

Haptic accessibility describes the condition in which interaction between a human and an artificial system becomes perceptually transparent. It occurs when the spatial, temporal, and inferential bandwidths of both systems are sufficiently aligned that tactile information flows bidirectionally without translation loss. In accessible systems, force, deformation, and context cues are rendered in the same representational structure that the human nervous system naturally expects, while the artificial system interprets human tactile behavior using comparable perceptual codes. Accessibility therefore marks the dissolution of the interface as a cognitive burden and the emergence of an embodied partnership between biological and artificial systems.

Haptic accessibility is the condition in which human and artificial haptic systems share sufficient parity in sensing, actuation, and inferential representation to enable perceptually transparent interaction. It arises when tactile information can pass bidirectionally between biological and artificial systems without significant loss of biomechanical, temporal, or perceptual fidelity. Achieving haptic accessibility reduces the informational and cognitive distance between user and system, allowing touch to function as a shared medium of perception and control rather than a constrained channel of communication.

7.3.1 Parity as the First Stage

Achieving parity is the immediate engineering objective of integrative haptics. The goal is not to replicate every nuance of the human skin but to reproduce the functional relationships that make human touch efficient and intelligible. Artificial skins with millimeter-scale spatial resolution, low-latency response, and inference-driven feedback loops can approach the perceptual thresholds that govern biological touch. In doing so, they enable operators to control robots, prosthetics, or virtual tools with a sense of continuity rather than mediation. Parity thus closes the accessibility gap, allowing information to pass through the interface as if it were a natural extension of the body.

Chapters 4–6 established the necessary preconditions for parity: inference-centric hardware, perceptually grounded software, and application-driven design goals. Parity draws these principles together by aligning the physical and informational structures of artificial systems with the natural constraints of human touch. A device achieves parity when its spatial resolution, mechanical compliance, and feedback latency fall within the psychophysical thresholds identified in Chap. 2—those defining the smallest perceivable changes in pressure, indentation depth, and temporal rate.

In the inspection gripper case study, for example, parity was approached by tuning tactile sensor arrays to replicate the deformation patterns that a human fingertip would experience when grasping materials of varying stiffness. The gripper's ability to estimate compliance from contact-area expansion and shear rate reflected not mere measurement but perceptual equivalence: its data streams could be mapped directly to human tactile cues without translation. Such systems demonstrate that perceptual realism emerges not from mimicking biology wholesale but from replicating the relationships among mechanical, neural, and inferential variables that make human touch intelligible. Parity, therefore, is the engineering manifestation of the first foundational principle—biomechanical–informational coupling—extended into the artificial domain.

7.3.2 *Augmentation as the Second Stage*

Where parity closes the perceptual gap, augmentation re-opens it as a frontier of exploration. Once systems achieve informational symmetry, their design can explore non-biological regimes of tactile experience: frequency ranges beyond human sensitivity, multi-point and multi-scale feedback, or persistent tactile memory that outlasts the physical event. Such augmentation transforms haptics from a medium of imitation into a medium of discovery, where new forms of perception and control become possible. Rather than merely human-like, these systems become human-aligned—anchored in psychophysical principles yet unconstrained by biological limits.

Once tactile information can circulate freely between human and artificial systems, design can pursue capabilities beyond biological limits. Chapter 5's discussion of multimodal integration hinted at this trajectory: software subsystems capable of predictive coding and generative modeling that can synthesize tactile experiences that have no direct physiological counterpart, such as anticipatory feedback or context-dependent tactile cues. Augmentation leverages these computational capabilities to extend both the dynamic range and the interpretive depth of touch.

Consider again the healthcare robot scenario from Chap. 6. Parity ensured that the robot could perceive human contact with appropriate sensitivity and emotional safety. Augmentation would allow it to sense micro-vibrations or thermal gradients imperceptible to humans, translating them into adaptive grip or comfort responses. Likewise, in teleoperation, augmented haptic channels could compress high-dimensional tactile data into intuitive perceptual codes, enabling operators to "feel" complex remote environments through simplified yet informative cues. These examples reflect the maturation of inference-centric design: the artificial system no longer imitates human touch but participates as a complementary perceptual partner. Augmentation thus reframes the goal of haptic technology from equivalence to reciprocity, where human and machine co-extend each other's sensory reach.

Haptic reciprocity represents the culminating state of integrative haptics—the point at which information exchange between human and artificial systems becomes not only symmetrical but mutually adaptive. In reciprocal systems, each side actively interprets, predicts, and responds to the other's tactile behavior within a shared inferential loop. The human no longer merely commands through touch, and the machine no longer merely reports it; both participate in an ongoing dialogue of sensing and meaning-making. Reciprocity therefore extends the notion of accessibility from perceptual transparency to co-perception: a condition in which biological and artificial agents collaborate through touch to construct a unified understanding of their shared environment. It is this mutual inference—rooted in parity but expanded through augmentation—that defines the final frontier of integrative haptics and transforms touch into a medium of partnership rather than control.

Haptic reciprocity is the condition in which human and artificial haptic systems engage in mutual inference—each interpreting and adapting to the tactile states, intentions, and feedback of the other within a shared closed-loop architecture. It extends haptic accessibility by enabling bidirectional adaptation, allowing touch to function not as a unidirectional command channel but as a medium of co-perception and cooperative understanding.

7.3.3 Universal Tactile Interaction

In the opening of this book, in Chap. 1, we described touch as "an intimate, dynamic dialogue between the body and the environment, shaping how we perceive and interact with everything around us." Haptic accessibility ultimately reframes human–machine interaction as a universal language problem rather than a hardware limitation. Touch, in both biological and artificial forms, is a communication channel with its own syntax and semantics—defined by pressure, deformation, and temporal rhythm. When the physical and inferential grammars of these systems align, interaction becomes fundamentally linguistic: contact encodes intent, resistance conveys acknowledgment, and force modulation expresses confidence or uncertainty. The integrated haptic system therefore functions not as an interface between discrete agents but as a shared medium of expression, enabling human and machine to negotiate meaning through embodied feedback. Such universality is not achieved by reproducing human sensation alone but by constructing a common representational substrate that both biological and artificial systems can inhabit. In this sense, the advancement of haptic technology parallels the evolution of natural language—it transforms coordination into communication and communication into shared cognition.

From this perspective, universal interaction becomes the defining socio-technical function of integrative haptics. As intelligent systems proliferate into industrial, domestic, and exploratory domains, their capacity to cooperate with humans depends on tactile fluency: the ability to read, predict, and respond through the same physical vocabulary that governs human action.

In the final frontier of haptic reciprocity, universal tactile interaction represents the culmination of integrative haptics: the emergence of a shared sensorimotor space in which human and artificial agents exchange information, intention, and adaptation through the medium of touch. It envisions a world where tactile communication operates as a foundational protocol—analogous to linguistic or digital languages—linking diverse substrates of intelligence. Within such a framework, contact is not merely a mechanical event but a bidirectional act of cognition, enabling coordination, learning, and reasoning to occur across biological and artificial boundaries. Universal tactile interaction thus defines the endpoint of the haptic continuum: from sensing and inference to accessibility and reciprocity, culminating in a unified architecture of embodied understanding.

> **Universal tactile interaction** is the condition in which biological and artificial systems share a common sensorimotor framework for exchanging information, intent, and adaptive feedback through touch. It unifies sensing, inference, and actuation across heterogeneous agents, enabling tactile communication to serve as a universal medium of coordination and learning within complex socio-technical environments.

In the end, haptic reciprocity transforms universal interaction from a design objective into a cognitive principle: the recognition that understanding itself can be mediated through contact. Within this framework, integrative haptics ceases to be the design of devices and becomes the architecture of shared perception—a foundation for how humans and machines will come to know the world together through touch.

7.4 Conclusion: Toward Reciprocal Tactile Intelligence

Haptic accessibility completes the intellectual arc of this book. In Section 7.1, integration establishes how systems achieve coherence; in Section 7.2, the information bottleneck defines why coherence is difficult to realize; in Section 7.3 accessibility demonstrates what lies beyond—reciprocal perception between human and machine. The long-term vision of integrative haptics is therefore not the replication of the human sense of touch but the creation of reciprocal tactile intelligences, each capable of sensing, interpreting, and responding within a shared inferential framework. In closing the gap between biological and artificial touch, we open a space for new kinds of embodied understanding—where interaction itself becomes a form of co-perception. In that shared tactile space—where perception becomes a dialogue—the future of haptics merges with the future of understanding itself.

Glossary

Active touch A perceptual process wherein an individual voluntarily controls the movement of their hand or other perceptual organs to gather tactile information from the environment or an object.

Afferent density The number of tactile nerve fibers (afferents) per unit area of skin.

Biomechanical-Informational Coupling The principle that tactile perception is both physical and informational. The physical arises from the biomechanical response of the skin to external forces; the informational arises from neural encoding and the pathways that generate psychophysical response. The relationship between biomechanics and psychophysics can be characterized through measurable mechanical cues that link physical deformation to perceptual experience.

Biomechanics The study of the structure, function, and motion of the mechanical aspects of biological systems. In haptics, it is the understanding of how forces and deformations are transmitted through the body's tissues (e.g., skin, muscles, tendons) during interaction with objects.

Central nervous system The part of the nervous system that integrates information received from the entire body and coordinates activity. It comprises the brain and spinal cord.

Cognitive load The total amount of mental effort being used in working memory. In haptic interactions, it is the cognitive resources allocated to processing tactile and force feedback.

Compliance A material property that measures the deformation of an object in response to an applied force.

Compliance discrimination The psychophysical ability of an individual to detect a difference in compliance between two objects.

Compliance perception The integrated, subjective experience of an object's compliance, resulting from the sensorimotor process of active or passive touch.

B. Li, T. Cody, *Design Principles for Integrated Haptic Systems*, Springer Series on Touch and Haptic Systems, https://doi.org/10.1007/978-3-032-16986-0

Cortex The outer layer of neural tissue of the cerebrum, responsible for higher-order brain functions including sensory perception, motor command generation, and cognition. In haptics, the primary somatosensory cortex (S1) and secondary somatosensory cortex (S2) are critical for processing tactile and proprioceptive information to form the percept of touch.

Cortical modulation The top-down, dynamic regulation of neural activity in sensory cortices by cognitive factors such as attention, expectation, and task context. In haptics, cortical modulation explains how the same peripheral input can be processed differently.

Cortical processing The multi-stage neural operations performed by the cerebral cortex on sensory input to transform it into a perceptual experience. For haptics, this involves the hierarchical integration of cutaneous and proprioceptive signals from the thalamus through primary (S1) and secondary (S2) somatosensory cortices, and association areas, to extract and represent features like texture, shape, and compliance.

Creep The time-dependent, continuous deformation of a material under a constant load, a key characteristic of viscoelastic substances like skin and soft tissues.

Cross-talk Unintended stimulation or interference between adjacent tactile actuators or sensory channels.

Cues The distinct physical signals that convey information about an object's properties during haptic interaction.

Cutaneous Pertaining to or affecting the skin. In haptics, "cutaneous cues" refer to the tactile sensations originating from the stimulation of receptors in the skin, distinct from kinesthetic cues related to limb position and movement.

Direct touch A mode of haptic interaction in which the user's body (typically the fingers) is in physical contact with a real object or surface.

Discrimination sensitivity A measure of the acuity of the haptic system, quantifying the smallest detectable difference between two stimuli. It is often measured as a Just-Noticeable Difference (JND) or a difference threshold.

Discrimination tasks A class of psychophysical experiments used to measure haptic perception, where an observer is required to identify whether two stimuli are the same or different, or to identify which of two stimuli has a greater magnitude on a specific dimension (e.g., roughness, stiffness). These tasks are fundamental for establishing discrimination sensitivities.

Discriminative touch The aspect of the tactile sense concerned with the perception and discrimination of spatial and temporal patterns on the skin.

Distributed actuation A design principle for haptic interfaces that employs an array of individual actuators distributed across a surface.

Elastic modulus A material property that quantifies its stiffness or resistance to elastic deformation under an applied load.

Elongation The deformation of a material characterized by an increase in length, typically expressed as a strain (change in length per unit original length).

Embodiment The sense that a tool or an external object (including a virtual body part) is integrated with one's own body schema and is under voluntary control.

Finite element models Computational models that simulate the mechanical behavior of complex structures by dividing them into a finite number of elements.

Force-feedback A class of haptic technology that provides users with kinesthetic information by generating resistive forces and torques.

Generative modeling A computational approach in which a model is trained to learn the underlying distribution of a dataset and can then generate new, synthetic data instances.

Glabrous skin The smooth, hairless skin found on the palmar surface of the hands and the plantar surface of the feet.

Grating orientation A psychophysical task used to assess the spatial acuity of the tactile system, wherein a participant identifies the orientation (e.g., parallel or perpendicular to the finger) of a grooved pattern pressed onto the skin.

Haptic accessibility The condition of perceptual transparency in human-artifact interaction, achieved when the spatial, temporal, and inferential bandwidths of the human and artificial haptic systems are sufficiently aligned. This alignment allows tactile information to flow bidirectionally without significant translation loss, creating a seamless and intuitive interface.

Haptic augmentation The design paradigm that explores tactile experiences beyond biological limits, enabled by achieving a foundation of haptic parity. It deliberately re-opens the perceptual gap to create non-biological regimes of touch, such as extended frequency sensitivity, multi-point feedback, or persistent tactile memory. This transforms haptics from a medium of imitation into a medium of discovery, creating human-aligned systems that are anchored in psychophysical principles but unconstrained by biology.

Haptic capability The functional capability of a haptic system, whether biological or technological.

Haptic hardware The physical components of a haptic system responsible for generating mechanical stimuli and/or measuring user input. This includes actuators, sensors, and the mechanical linkages that constitute the interface between the user and the machine.

Haptic information bottleneck A fundamental limitation in haptic technology where the device's capability to transmit information to the user is significantly lower than the human haptic system's capacity to receive it.

Haptic parity The engineering objective of closing the perceptual gap between natural and artificial touch. It aims not to replicate the full biological nuance of the skin, but to reproduce the functional relationships that make human touch efficient and intelligible. Parity is achieved when the artifact's haptic capabilities are sufficiently aligned with the human haptic system's bandwidths to enable effective interaction.

Haptic reciprocity The culminating state of integrative haptics, representing a mutually adaptive information exchange between human and artificial systems. In reciprocal systems, information flow is not only symmetrical but involves each side actively interpreting, predicting, and responding to the other's tactile behavior within a shared inferential loop, enabling a deeply collaborative interaction.

Haptic science The interdisciplinary study of the sense of touch, encompassing the neurophysiology of the haptic system, the psychophysics of tactile and kinesthetic perception, and the biomechanics of physical interaction. It provides the foundational knowledge that informs the design and development of haptic technology.

Haptic system A functional system composed of interacting components that work together to deliver haptic capability, to support perception and interaction within a larger physical or digital architecture.

Haptic system design goal An intended objective or desired quality that guides the development of a haptic system to ensure it meets functional requirements and performs effectively within its context.

Haptics The science and engineering of the sense of touch involving sensing and action to perceive, interpret, and respond to object and spatial properties.

Hierarchical modulation The neural process by which higher-level brain regions exert top-down control over lower-level sensory processing regions. This modulation, driven by attention, expectation, or task goals, shapes the neural representation of haptic information as it flows through the cortical hierarchy.

Incipient slip The initial, localized micro-slip that occurs at the periphery of a contact region just before full slip.

Indentation depth The perpendicular displacement of a contactor or object into the skin surface.

Indirect touch A haptic interaction paradigm where the user manipulates a tool or interface to explore an environment, rather than using direct finger contact.

Individual Variability The principle that the differences in haptic perception and discrimination performance among individuals, arising from factors such as age, sex, finger size, skin properties, and prior experience.

Inference-centric haptic systems An integrated haptic system in which coordination across hardware, software, and application subsystems enables the transformation of tactile signals into perceptual inferences that guide action and decision-making.

Integrated haptic system A haptic system composed of three coordinated subsystems—hardware, software, and application—engineered to deliver haptic capabilities through closed-loop interaction.

Integrated haptic system design process A four-step method for designing integrated haptic systems: (1) identify haptic system design goals from system design goals; (2) identify relevant first principles; (3) decompose goals across hardware, software, and application subsystems guided by those principles; and (4) design, develop, and iterate to achieve system-level goals.

Integrative haptics The interdisciplinary science and engineering of touch that unifies biomechanical sensing, computational inference, and system-level application within a closed-loop architecture. It treats haptic capability as an emergent property of coordinated hardware, software, and behavioral subsystems designed to achieve perceptually coherent, adaptive, and context-aware interaction.

Kinesthetic Pertaining to the perception of body position, movement, and forces, mediated by receptors in the muscles, tendons, and joints.

Mechanical cues The physical signals arising from object interaction that are transduced by the haptic system, including forces, vibrations, skin deformations, and temperature changes.

Mechanoreceptor density The number of tactile nerve endings per unit area of skin. This density, which is highest in the fingertips, determines the spatial resolution of the haptic system.

Mechanoreceptors Specialized sensory nerve endings in the skin that respond to mechanical stimuli such as pressure, vibration, and stretch. The four major types in glabrous skin are Merkel disks, Meissner's corpuscles, Ruffini endings, and Pacinian corpuscles, each with distinct response properties.

Mediated touch A broad term for any haptic interaction that is facilitated by a technology, encompassing both indirect touch via a tool and touch through a haptic interface. It contrasts with direct touch, where the finger makes unassisted contact with a physical object.

Micro-slip Small, localized relative motion between the skin and a contacted object that occurs within the overall area of contact.

Microneurography A neurophysiological technique in which a fine micro-electrode is inserted into a peripheral nerve in an awake human to record neural activity from single tactile afferent fibers. It has been fundamental for establishing the relationship between mechanical stimuli and the responses of individual mechanoreceptors.

Modular software A software architecture for haptic systems where functionality is separated into independent, interchangeable modules. This promotes flexibility, reusability, and easier maintenance of complex haptic applications.

Modulation of skin mechanics The alteration of the skin's mechanical properties. These changes directly affect how contact forces are transformed into skin deformations, thereby influencing the input to the mechanoreceptors and the resulting tactile perception.

Multi-axial sensing The ability to detect forces, torques, and moments along or around multiple axes.

Multi-channel Pertaining to a system that uses multiple, independent information pathways.

Multifunctional Capability of rendering more than one type of tactile cue (e.g., vibration, pressure, and shear) or simulating multiple object properties.

Multisensory convergence The integration of information from multiple sensory modalities (e.g., vision, audition, haptics) within the brain to form a unified percept.

Neural encoding The process by which physical stimuli are transformed into patterns of neural activity in sensory neurons.

Neural firing The generation and propagation of an action potential (or neural spike) along an axon. This is the fundamental unit of communication in the nervous system, and the pattern of firing in tactile afferents conveys information about a stimulus to the brain.

Neural firing patterns The temporal sequence of action potentials generated by a neuron or population of neurons. Different stimulus features are represented by distinct firing patterns.

Neural firing rates The frequency at which a neuron generates action potentials, typically measured in spikes per second.

Neural spike latency The time delay between the onset of a stimulus and the generation of the first action potential by a sensory neuron.

Neural spikes The all-or-none electrochemical impulses (action potentials) that are the primary means of communication along nerve fibers. The timing and sequence of these spikes in tactile afferent nerves constitute the neural representation of a haptic stimulus.

Neurophysiology The branch of physiology that studies the function of the nervous system. It provides the foundation for understanding how mechanoreceptors transduce mechanical signals and how the central nervous system processes this information to create the sense of touch.

Nociceptors Sensory receptors that respond to potentially damaging stimuli, resulting in the perception of pain. They are critical for simulating safety-critical interactions and protective sensory feedback.

Optical-based sensors A class of tactile sensors that use light to measure contact parameters. They are valued for their high sensitivity, immunity to electromagnetic interference, and potential for high spatial resolution.

Partial slip A contact state in which a portion of the skin-object interface undergoes localized micro-slip while the remainder remains stuck. This transitional phase generates distinct vibrational cues that are critical for the reflexive adjustment of grip force to prevent full object slip.

Passive touch A perceptual mode where tactile stimuli are applied to a stationary, receptive skin surface.

Peripheral encoding The process by which the physical properties of a tactile stimulus are transduced into a sequence of neural action potentials by mechanoreceptors in the skin. This initial stage of sensory processing determines the information available to the central nervous system.

Peripheral mechanoreceptors Specialized sensory nerve endings located in the skin, muscles, and tendons that transduce mechanical energy into neural signals.

Proprioception The sense of the relative position and movement of one's own body parts, and the forces exerted by the body. It is mediated by receptors in the muscles, tendons, and joints.

Psychophysics The scientific discipline concerned with the quantitative relationship between physical stimuli and the sensations and perceptions they evoke.

Receptive field The region of the skin surface within which a mechanical stimulus will alter the firing rate of a particular sensory neuron. The size and structure of a neuron's receptive field determine its spatial resolution.

Sensory organs The biological structures dedicated to transducing specific forms of environmental energy into neural signals. For haptics, the primary sensory organs include the skin (for cutaneous touch) and the receptors in the muscles, tendons, and joints (for proprioception).

Signal-to-noise A measure that compares the level of a desired signal to the level of background noise.

Skin deformation The change in geometry of the skin surface and subsurface layers resulting from physical contact with an object.

Skin mechanics The physical properties and deformation behavior of the skin under mechanical load, including its viscoelasticity, anisotropy, and nonlinearity.

Skin microstructure The small-scale anatomical features of the skin, including the epidermis, dermis, fingerprints (epidermal ridges), sweat ducts, and hair follicles.

Skin properties The intrinsic mechanical characteristics of skin that define its response to external stimuli.

Skin state The dynamic physiological condition of the skin at a given moment.

Slip detection The perceptual or sensor-based process of identifying the onset of relative motion between the skin and a contacted object.

Slip onset The initial moment when relative motion begins at the skin-object interface.

Social touch A modality of touch concerned with affective and communicative interactions between individuals. It is mediated by a distinct class of unmyelinated nerve fibers (C-tactile afferents).

Somatosensory inhibition A neural process whereby the activation of one somatosensory pathway suppresses the activity of another. This can occur spatially (lateral inhibition) to sharpen perceptual localization, or temporally to filter out irrelevant signals, enhancing the signal-to-noise ratio of tactile perception.

Somatosensory system The comprehensive network responsible for bodily sensation, encompassing the sensory receptors, neural pathways, and brain regions that process touch, temperature, pain, and proprioception.

Spatial acuity The ability to resolve fine spatial details. It is largely determined by mechanoreceptor density and innervation patterns.

Systems engineering An interdisciplinary approach to designing and managing complex systems over their life cycles centered on the concept that systems are the whole formed by interrelated parts.

Tactile acuity Synonymous with spatial acuity, it is the capacity for fine spatial discrimination using the sense of touch. It is commonly measured using the two-point discrimination test or grating orientation task.

Tactile discrimination The perceptual ability to distinguish between two or more tactile stimuli based on differences in their properties.

Tactile perception The cognitive and neural process of interpreting and giving meaning to sensory information derived from the stimulation of mechanoreceptors in the skin. It is the subjective experience resulting from the transduction, encoding, and processing of tactile signals.

Tactile sensitivity The ability to detect the presence of a tactile stimulus.

Two-point spatial discrimination test A classic psychophysical measure of tactile spatial acuity. It determines the smallest distance between two simultaneously applied points of contact at which they can be perceived as distinct rather than a single point.

Vibrotactile feedback A method of haptic stimulation that uses controlled vibrations to convey information.

Viscoelastic A material property describing a substance that exhibits both viscous (liquid-like, time-dependent) and elastic (solid-like, instantaneous) characteristics when undergoing deformation. Skin and many soft biological tissues are viscoelastic, meaning their response to force depends on both the magnitude and the rate of loading.

Weber fraction A key psychophysical constant that quantifies the just-noticeable difference (JND) for a stimulus. A smaller Weber fraction indicates greater sensitivity to change.

Index

B. Li, T. Cody, *Design Principles for Integrated Haptic Systems*, Springer Series on Touch and Haptic Systems, https://doi.org/10.1007/978-3-032-16986-0

The manufacturer's authorised representative in the EU is Springer Nature Customer Service Centre GmbH, Europaplatz 3, 69115 Heidelberg, Germany. If you have any concerns regarding our products, please contact ProductSafety@springernature.com

Printed and bound by CPI Group (UK) Ltd, Croydon, CR0 4YY
05/10/2026
02228059-0007